Studies in Computational Intelligence

Volume 1203

The series "Studies in Computational Intelligence" (SCI) publishes new developments and advances in the various areas of computational intelligence—quickly and with a high quality. The intent is to cover the theory, applications, and design methods of computational intelligence, as embedded in the fields of engineering, computer science, physics and life sciences, as well as the methodologies behind them. The series contains monographs, lecture notes and edited volumes in computational intelligence spanning the areas of neural networks, connectionist systems, genetic algorithms, evolutionary computation, artificial intelligence, cellular automata, self-organizing systems, soft computing, fuzzy systems, and hybrid intelligent systems. Of particular value to both the contributors and the readership are the short publication timeframe and the world-wide distribution, which enable both wide and rapid dissemination of research output.

Indexed by SCOPUS, DBLP, WTI AG (Switzerland), zbMATH, SCImago.

All books published in the series are submitted for consideration in Web of Science.

Nikolaos Polatidis · Elias Pimenidis ·
Marcello Trovati · Mehmet E. Aydin
Editors

Intelligent Distributed Computing XVII

17th International Symposium on Intelligent Distributed Computing, IDC 2024

Springer

Editors
Nikolaos Polatidis
School of Architecture, Technology and Engineering
University of Brighton
Brighton, UK

Elias Pimenidis
School of Computing and Creative Technologies
University of the West of England
Bristol, UK

Marcello Trovati
Department of Computer Science
Edge Hill University
Ormskirk, UK

Mehmet E. Aydin
School of Computing and Creative Technologies
University of the West of England
Bristol, UK

ISSN 1860-949X ISSN 1860-9503 (electronic)
Studies in Computational Intelligence
ISBN 978-3-031-87638-7 ISBN 978-3-031-87639-4 (eBook)
https://doi.org/10.1007/978-3-031-87639-4

This Springer imprint is published by the registered company Springer Nature Switzerland AG
The registered company address is: Gewerbestrasse 11, 6330 Cham, Switzerland

Editor Note

The 17th Annual International Symposium on Intelligent Distributed Systems (IDC2024) was hosted by the University of Brighton, UK, 18–19 September 2024.

This volume includes all 16 full papers that were accepted for presentation after a rigorous peer review process by experts in their respective fields. The papers are presented here grouped in five areas of focus, namely *Distributed Systems*, *Cyber Security*, *Predictive Modelling*, *Blockchain Applications*, and *Large Language Models (LLMs)*. They provide a wide range of areas of application that demonstrate the dynamics, and constantly evolving nature of Distributed Systems research.

The authors work in, two continents, six countries, and a total of 14 different institutions and organizations.

The work spans different levels of progress of research, completed projects at institutional level, outcomes of Ph.D. student work, and results that relate to large transnational projects funded by the European Union and other national and international funders.

We hope that the readers of this volume of proceedings will find the contents informative and an opportunity to engage with new outcomes of innovative research.

Nikolaos Polatidis
Elias Pimenidis
Marcello Trovati
Mehmet E. Aydin

Contents

Predictive Modelling

Blockchain Applications

Large Language Models (LLMs)

Distributed Systems

NeuroEvolution Algorithms Applied in the Designing Process of Biohybrid Actuators

Hugo Alcaraz-Herrera, Michail-Antisthenis Tsompanas, Igor Balaz, and Andrew Adamatzky

Abstract Soft robots diverge from traditional rigid robotics, offering unique advantages in adaptability, safety, and human-robot interaction. In some cases, soft robots can be powered by biohybrid actuators and the design process of these systems is far from straightforward. We analyse here two algorithms that may assist the design of these systems, namely, NEAT (NeuroEvolution of Augmented Topologies) and HyperNEAT (Hypercube-based NeuroEvolution of Augmented Topologies). These algorithms exploit the evolution of the structure of actuators encoded through neural networks. To evaluate these algorithms, we compare them with a similar approach using the Age Fitness Pareto Optimisation (AFPO) algorithm, with a focus on assessing the maximum displacement achieved by the discovered biohybrid morphologies. Additionally, we investigate the effects of optimisation against both the volume of these morphologies and the distance they can cover. To further accelerate the computational process, the proposed methodology is implemented in a client-server setting; so, the most demanding calculations can be executed on specialised and efficient hardware. The results indicate that the HyperNEAT-based approach excels in identifying morphologies with minimal volumes that still achieve satisfactory displacement targets.

Keywords Neuroevolution · NEAT · HyperNEAT · Biohybrid machines · Client-server model · Optimisation

H. Alcaraz-Herrera (✉) · M.-A. Tsompanas · A. Adamatzky
Unconventional Computing Laboratory, University of the West of England, Bristol, UK
e-mail: hugo.alcaraz@uwe.ac.uk

M.-A. Tsompanas
School of Computing & Creative Technologies, University of the West of England, Bristol, UK

I. Balaz
Laboratory for Meteorology, Physics and Biophysics, Faculty of Agriculture, University of Novi Sad, Novi Sad, Serbia

N. Polatidis et al. (eds.), *Intelligent Distributed Computing XVII*, Studies in Computational Intelligence 1203, https://doi.org/10.1007/978-3-031-87639-4_1

1 Introduction

The primary focus of the robotics field is designing and building machines capable of assisting humans across a variety of tasks such as surgery [15], agriculture [18], and manufacturing [3]. Soft robots are a sub-category utilising flexible, deformable materials [9, 10, 20, 31], as opposed to traditional rigid robots. These machines mimic the movement and behaviour of living organisms, making them better-suited for various applications, particularly in fields like healthcare, search and rescue, and space exploration.

Some notable examples of soft robotics based on biological inspiration for their design have been proposed in the past [28]. For instance, developments of soft robotic suckers inspired by octopus tentacles have been studied [8]. This highlights the use of flexible materials and smart adhesion. Another study demonstrated a 3D-printed soft robot powered by a combustion reaction [2], that showcased the use of innovative materials and manufacturing techniques. Also, a soft robot capable of navigating its environment by growing, in a similar way as a plant was implemented [4]. There are several original ideas and implementations on plant roots-based inspiration for soft robots [11, 21].

The field of soft robotics continues to advance, with researchers drawing inspiration from biology to create robots that can interact with delicate environments and objects more effectively. As the field matures, it holds great promise for a wide range of applications, and ongoing research continues to push the boundaries of what soft robots can achieve. More recently, robotics not only take inspiration from biology, but even employ biological tissue as part of their functionality [12, 13].

Given the challenges that arise from utilising living tissues and the attempt to control them, automated techniques for designing soft robots were studied to unveil the capacity of these novel machines to tackle specific problems. The methodology presented in our current work represents the initial steps towards the development of a biohybrid machine (BHM) actuator that will be utilised for biomedical applications.

Typical robot design is a complex process, since it may imply numerous iterations of testing designs in real life to alleviate the effect of the reality gap in simulations. As a consequence, previous studies required substantial time and material resources to be spent to find the most suitable robot design [23]. For soft robots, challenges are even more complex [5]. The added complexity primarily arises from the use of flexible materials, characterised by intricate and nonlinear mechanical properties. BHMs integrate biological components, such as cells or tissues, with synthetic elements, a fact that introduces an additional layer of complexity. These BHMs effectively merge the unique capabilities of biological systems with the robust characteristics offered by synthetic materials. However, this amalgamation also adds to the already intricate design landscape by introducing a multitude of parameters that must be carefully considered.

To tackle these challenges that arise during robot design considerations, an interesting approach was proposed, namely NeuroEvolution (NE). NE involves the evolution of topologies and connection weights of artificial neural networks (ANNs) employing a genetic algorithm (GA). Arguably, the most prominent approach to evolve ANNs is NeuroEvolution of Augmenting Topologies (NEAT) [26], which has been successfully applied in several fields of robotics such as morphology design [1] and gait generation [16]. Moreover, NEAT was extended in order to reproduce natural patterns in the learning process. This new approach, called HyperNEAT, employs a type of ANNs called Compositional Pattern-Producing Networks (CPPNs) to generate symmetry and other patterns in the evolved topologies [25].

The primary aim of our research is to evaluate the applicability of NEAT and HyperNEAT on the BHM design process. We assess the performance of the methodology utilising NEAT and HyperNEAT in comparison to an approach based on Age Fitness Pareto Optimisation (AFPO) [22]. The comparison is based on two key metrics of the produced BHM morphologies, namely the maximum displacement achieved within a specified time-frame and the trade-off between the total volume of the morphologies and the displacement attained.

Furthermore, the implementation of the proposed methodology that utilises NEAT and HyperNEAT has been designed by taking into consideration multiprocessing as a way to accelerate computations. More specific, the implementation is following a client-server model to outsource the more computationally intensive part of the optimisation process on distributed and more efficient computing platforms.

2 Background

NEAT has been widely used in robotics, particularly in locomotion tasks. For instance, NEAT alongside CPPNs were utilised to develop a model that improves the locomotion of snake-like modular robots [24]. That study compared the performance of NEAT against a Multiobjective Genetic Algorithm (MOGA) in three different environments. Results showed that NEAT outperforms MOGA in locomotion control throughout all the predefined environments. Under the same paradigm, *Combined-CPPN-NEAT* [6] was introduced as a method to design robots capable of performing multiple tasks. This method integrated the topologies of networks that were first evolved to solve single tasks. The method was compared against a MOGA using NEAT in two scenarios: (i) designing morphologies, and (ii) locomotion tasks. Results indicate that Combined-NEAT-CPPN was able to overcome MOGA-NEAT in both scenarios.

Taking into account works of simulated-only robotics, NEAT has been applied to guide the designing process and the control of robots. For example, the performance of NEAT against the Accuracy-Based Learning Classifier System (XCS) was compared in Robocode, a simulator of tanks [14]. Each tank was driven by a Java agent, which had an extensive range of behaviours and sensory data. Both approaches were compared in scanning and targeting tasks. NEAT demonstrated a suitable performance; however, it required an extensive training period. Also, XCS could learn fast, except in skewed payoff scenarios. Furthermore, under this domain, both approaches were susceptible to over-fitting. In the same direction, NEAT was applied to develop locomotive behaviours of virtual creatures [27]. This study utilised a physics engine called *Open, Dynamic Engine* (ODE) for experimentation. NEAT was compared against a more traditional approach to evolve ANNs [19]. Three behaviours were tested: crawling, running, and somersault. Results indicate that the behaviours generated by NEAT tend to be more robust, and they take better advantage of the capacity of the morphologies than those generated by the traditional approach.

3 Methods

The NE algorithms utilised here are population-based optimisation methods following GA methodology. Individuals of the populations are networks (ANNs and CPPNs) that encode 3D BHM morphologies. The 3D morphologies are evaluated by a physics engine called *Voxelyze* [5]. The output of the physics engine used here consists of: (a) the absolute displacement of the morphology from its initial position during the lapse set (10 s); and (b) the number of voxels that compose the morphology. Voxels are the elementary building blocks in Voxelyze that can simulate different materials, i.e., active or passive with different parameters (details on the utilisation of the simulator are provided in previous works [7, 30]).

HyperNEAT evolves CPPNs that in turn design ANNs [25]. Thus, based on a three-dimensional layout (x, y, z) where BHM morphologies need to be designed, the number of input neurons of the substrate is three. Substrate defines the ANNs that will decode into 3D BHM morphologies. Furthermore, it is necessary to determine the presence or not of a voxel for each position (i.e., a unique (x,y,z) tuple) of the layout and, if present, the type of material the voxel is made of (i.e., active or passive). Thus, the number of output neurons of the substrate is two. Figure 1 depicts the neuron allocation of the substrate utilised during experimentation.

Based on the aforementioned definition, the substrate is queried as follows:

$$SUBSTRATE(x_i, y_i, z_i) = PV_i, M_i \tag{1}$$

where $[x_i, y_i, z_i]$ represent the coordinates of the i point in the layout, whereas PV_i represents the presence of a voxel in the i point of the layout, and M_i refers to the type of material the voxel is made of.

It is essential to emphasise that the neuron allocation used for these experiments was chosen after a series of experiments on various numbers of neurons in the $[1, 7]$ range per layer and various numbers of layers in the $[1, 7]$ range. Regarding the activation function implemented in the substrate, *ReLU* was chosen after experimenting with a set of activation functions.

Note here that the NEAT mechanism does not require a substrate to operate. On the contrary, NEAT applies GA operations on the CPPNs that will directly encode the 3D morphologies. When applying NEAT, the number of dimensions in the layout where BHM morphologies are designed determines the number of inputs for CPPNs. In this case, the environment is three-dimensional, so, the number of inputs is set to

Fig. 1 Substrate utilised under HyperNEAT for morphology generator.

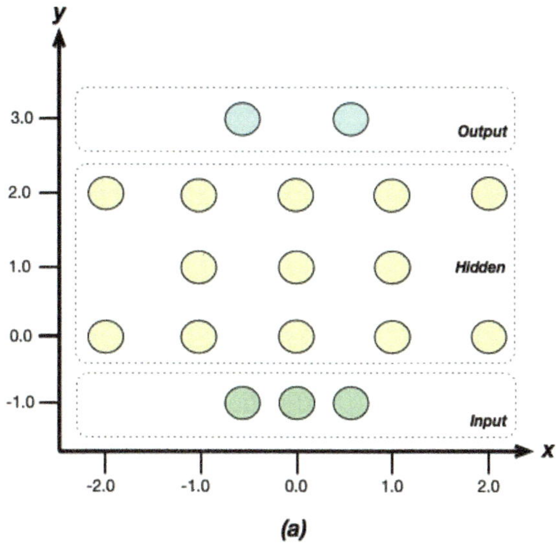

three. Additionally, based on the voxel encoding, the number of output neurons is set to two. Thus, CPPNs are queried as follows:

$$CPPN(x_i, y_i, z_i) = PV_i, M_i \tag{2}$$

where $[x_i, y_i, z_i]$ refer to the coordinates of the i point in the layout. Regarding PV_i, it represents the presence of a voxel in the i point of the layout. Furthermore, M_i represents the type of material the voxel is made of in the i point of the layout.

In order to analyse the performance of BHM morphologies generated by NEAT and HyperNEAT, two metrics are employed (i) the displacement of the BHM during a simulation of 10 s and (ii) the trade-off between displacement and the number of voxels within the simulated BHM (i.e. the total volume of the BHM). The outcomes of both metrics are studied under 500 different controller scenarios (i.e., each active voxel expands at a different phase offset that is generated at random for each scenario), in order to represent the inherent difficulties of controlling biological cells. Furthermore, an approach based on AFPO [7] is used to generate morphologies as a baseline metric to compare the results from NE algorithms.

Due to the considerable computation time required when running the physics engine to evaluate individuals, the code used as a baseline has been written under a multiprocessing paradigm [7]. That is a strategy to reduce the total execution time as much as possible. Our implementation utilising NEAT and HyperNEAT has also been designed considering multiprocessing as a technique to attenuate the time spent for evaluation. In specific, the software has been written under a client-server architecture to take advantage of distributed computing capacities. Figure 2 depicts the general workflow of the implementation. The evaluation of individuals can take

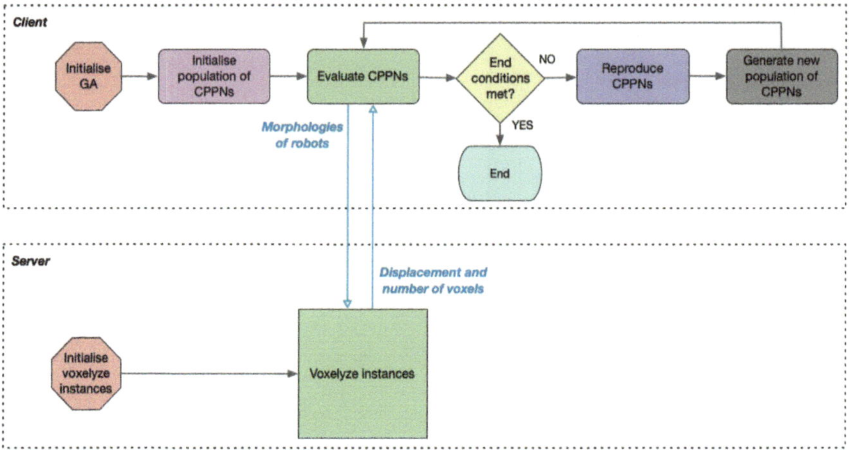

Fig. 2 Diagram of the implementation of NEAT and HyperNEAT for producing BHM morphologies

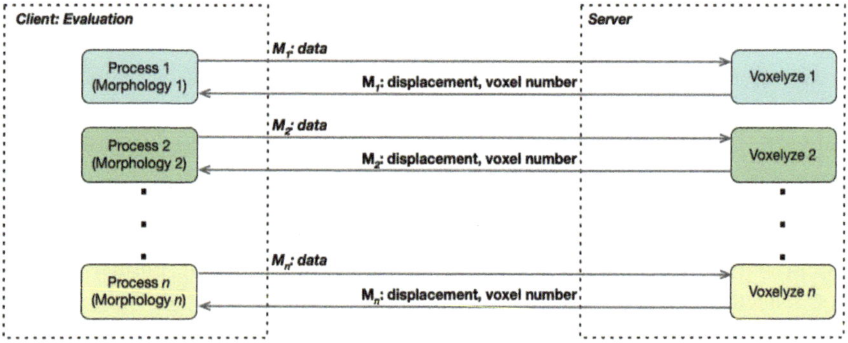

Fig. 3 Evaluation of individuals with processes triggered and concurrent interactions with Voxelyze instances

place concurrently, which leads to asynchronous execution. Figure 3 illustrates in more detail the evaluation stage. Thus, the data output of the server-side is utilised to calculate the fitness of morphologies. It is essential to point out that communication is established by HTTP, employing GET and POST methods. JSON is the format for transferring data between the NE GA (executed on the client-side) and the fitness function (executed on the server-side), which includes a reverse proxy acting as a load balancer.

Due to the flexible and scalable nature of the developed software, it is plausible to implement it in several diverse infrastructures. For instance, the server can be deployed in a virtual machine or any cloud computing platform. Regarding the client, it can be deployed regardless of the operative system and the architecture. Moreover, the software takes full advantage of computational power in both ends (i.e., the number of cores in the CPU). Finally, three virtual machines were used here during experimentation, one for each approach. The hardware configured for all virtual machines is described as follows:

- *Processor*: ARM, 9 cores (18 threads), 3.20 GHz.
- *RAM memory*: 16GB, LPDDR5.

3.1 Experimental Set-up

For all approaches, the population contains 50 encoded individuals (i.e., CPPNs or ANNs that will translate into 3D actuator morphologies), and each evolutionary run is terminated after 1000 generations. Moreover, the available activation functions are as follows: *sine*, *negative sine*, *absolute*, *negative absolute*, *square*, *negative square*, *square root of absolute*, *negative square root of absolute* and *sigmoid*. The implementation used to generate the AFPO-based morphologies is detailed in [7], available on GitHub[1] and it was not modified during experimentation. Under the three approaches, the population is initialised following the procedure described in [7] to provide a meaningful comparison of the outputs. Regarding NEAT and HyperNEAT, Table 1 presents the parameters used during the evolution of CPPNs. For each approach, five experimental evolution runs were conducted with different initial populations.

Table 1 Parameters utilised to evolve CPPNs under NEAT and HyperNEAT in the morphology generator domain

Parameter	Value
Compatibility threshold	3
Compatibility disjoint coefficient	1.0
Compatibility weight coefficient	0.5
Maximum stagnation	15
Survival threshold	0.3
Activation function mutate rate	0.4
Adding/deleting connection rate	0.3/0.2
Activating/deactivating connection rate	0.5
Adding/deleting node rate	0.3/0.2

[1] https://github.com/skriegman/reconfigurable_organisms.

4 Results

Since the purpose of the experiments described here is to test the suitability of morphologies generated by AFPO, NEAT and HyperNEAT in terms of locomotion capabilities and robustness, two series of experiments are conducted: (i) finding the most capable morphology in terms of reaching the maximum displacement possible; and (ii) identifying morphologies that reach the maximum displacement possible, in addition to containing the minimum number of voxels. Here the design space is defined by the x and y axis having a range of $[0, 8]$ voxels, while the z axis has a range of $[0, 7]$ voxels.

4.1 Maximising Displacement

One of the crucial features of BHM morphologies is the displacement capability, regardless of the external conditions and under a possibly loosely defined control strategy. Thus, the fittest morphologies generated by AFPO, NEAT and HyperNEAT provided after 1000 generations, were tested using 500 different phase offsets of each of the active voxels in a morphology. These controller phase offset scenarios were a priori generated (at random) and used consistently for all the morphologies generated by all approaches. It is noteworthy that this experiment does not consider the number of voxels that compose morphologies as a penalty to the fitness. For reference, the amount of voxels per morphology generated by the three methods are provided in Table 2.

Figure 4 presents violin plots comparing the performance in terms of displacement observed by morphologies generated by AFPO, NEAT, and HyperNEAT. Each violin shows the kernel density estimation of the frequency distribution, maximum, minimum and median of displacement values across the 500 controller runs (i.e., phase offset scenarios).

Under AFPO (see Fig. 4a), all morphologies present significant differences. The produced data are not normally distributed (Shapiro-Wilk test; $p < 0.01$). Then, by utilising the Kruskal-Wallis test, it is feasible to state that among the displacements observed, significant differences exist ($p < 0.01$). Derived from this result, it is feasible to rank the maximum displacement reached by the morphologies as in:

Table 2 Number of voxels of the five fittest morphologies generated by AFPO, NEAT, and HyperNEAT (total of 15 morphologies)

Approach	Morphology 1	Morphology 2	Morphology 3	Morphology 4	Morphology 5
AFPO	229	283	224	244	240
NEAT	192	192	128	128	224
HyperNEAT	118	122	165	218	116

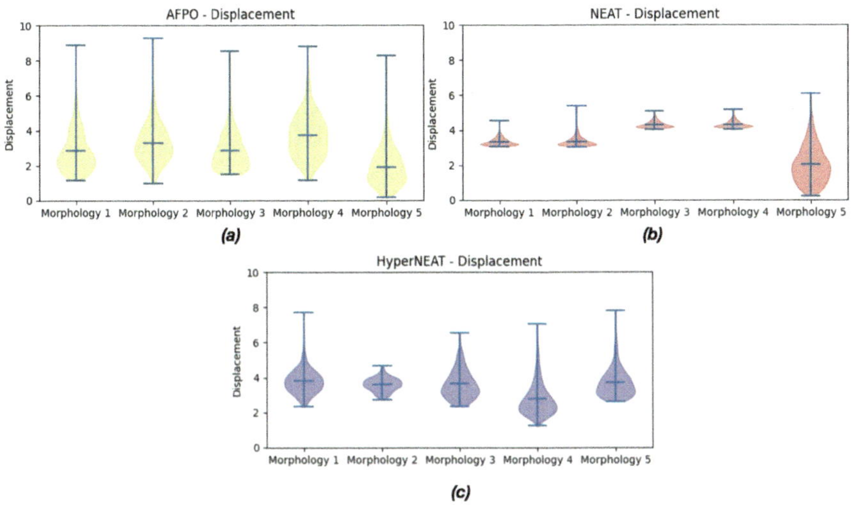

Fig. 4 Displacement observed under 500 different phase offset scenarios for the controllers of the fittest morphologies found in five different evolutionary runs under: **a** AFPO-based approach, **b** NEAT-based approach, and **c** HyperNEAT-based approach

Morphology 4 > Morphology 2 > Morphology 1 > Morphology 3 > Morphology 5 (Dunn's test: $p < 0.01$).

Regarding NEAT (see Fig. 4b), similar results are obtained; all morphologies exhibit significant differences. The data obtained are not normally distributed (Shapiro-Wilk test; $p < 0.01$). Then, by employing the Kruskal-Wallis test, it can be confirmed that among the displacements obtained, significant differences are emerging ($p < 0.01$). Thus, ranking the performance of the morphologies in terms of maximum displacement reached can be performed: Morphology 4 > Morphology 3 > Morphology 1 > Morphology 2 > Morphology 5 (Dunn's test: $p < 0.01$).

Under HyperNEAT all morphologies exhibit significant differences (see Fig. 4c). Again, all collected data are not normally distributed (Shapiro-Wilk test; $p < 0.01$). Then, employing the Kruskal-Wallis test, it is possible to confirm significant differences ($p < 0.01$). Based on this result, it is plausible to rank the performances (i.e., maximum displacement reached) of the morphologies generated by HyperNEAT: Morphology 1 > Morphology 5 > Morphology 3 > Morphology 2 > Morphology 4 (Dunn's test: $p < 0.01$).

Finally, Fig. 5 depicts violin plots comparing the one best performing morphology out of the fittest ones generated by AFPO, NEAT, and HyperNEAT in terms of reaching maximum displacement. Each violin shows the kernel density estimation of the frequency distribution, maximum, minimum and median of values across 500 runs (i.e., phase offset controller scenarios). The Kruskal-Wallis test makes it feasible to confirm the existence of significant differences among the fittest morphologies

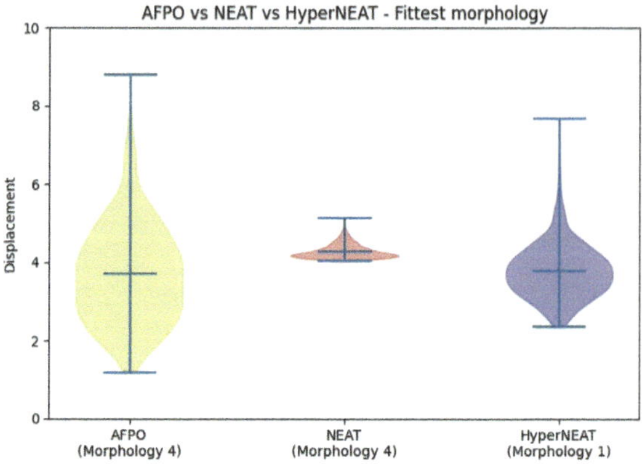

Fig. 5 Fittest morphology in terms of reaching maximum displacement under: AFPO, NEAT, and HyperNEAT

($p < 0.01$). Hence, the performance can be ranked as follows: NEAT > HyperNEAT > AFPO (Dunn's test: $p < 0.01$).

Consequently, the results suggest that, generally, NE-based approaches can produce compact morphologies that can reach higher displacement than those generated by AFPO, regardless of the phase offset scenario. This can lead to the conclusion that the morphologies produced by NE are more robust and require fewer voxels (see Table 2). Moreover, when NEAT is compared against HyperNEAT, it can generate morphologies that reach longer displacements and, considering the tight data distribution observed, their performance tends to be more consistent regardless of the controller scenario.

4.2 *Minimising Volume and Maximising Displacement*

Although the capability of displacement is crucial, additional aspects may be required to consider a morphology as suitable for production in real-life. Since morphologies represent BHMs that are aiming to be manufactured in minuscule scales (e.g., part of a catheter manoeuvring within human blood vessels), a morphology with fewer voxels (thus, of smaller volume) is considered more suitable and less energy consuming. In order to evaluate the suitability of morphologies, the evaluation will now consider: (a) their capacity on achieving displacement (maximising target); and (b) the amount of voxels comprising them (minimising target). The mechanism to evaluate the displacement of morphologies (δ) is defined as follows:

$$\delta_m = \frac{\Delta_m}{\Delta_{max}} \tag{3}$$

where Δ_m is the displacement of morphology m, which is obtained as an output of the simulation on the physics engine, and Δ_{max} is the maximum displacement configured

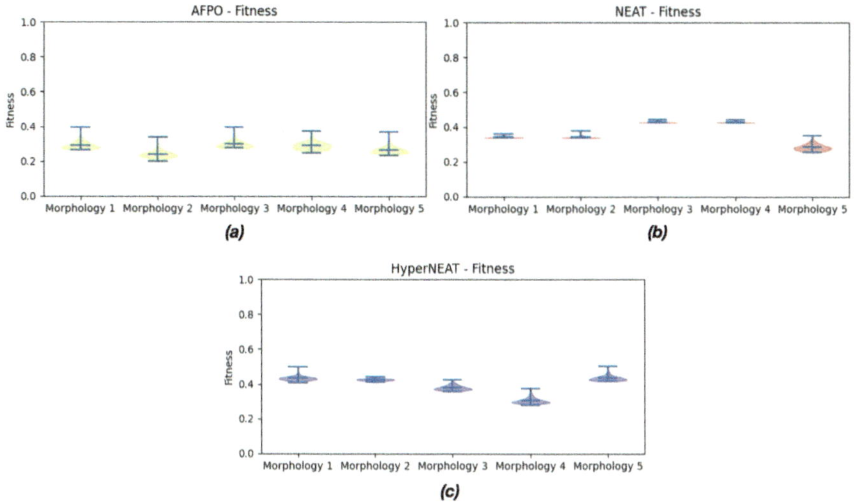

Fig. 6 Fitness measured under 500 phase offset scenarios for the controllers of the fittest morphologies found in five different evolutionary runs under: **a** AFPO-based approach, **b** NEAT-based approach, and **c** HyperNEAT-based approach.

for this experiment, which is set to 20. Furthermore, δ is normalised in the [0.0, 1.0] range. The evaluation of the number of voxels of a morphology (ν), is characterised as follows:

$$\nu_m = 1 - \frac{\Upsilon_m}{\Upsilon_{max}} \tag{4}$$

where Υ_m represent the number of voxels contained in morphology m and Υ_{max} is the maximum number of voxels possible, which is given by the dimension of each axis composing the layout where BHMs are designed: $8 \times 8 \times 7 = 448$; consequently, $\Upsilon_{max} = 448$. Moreover, ν is normalised in the [0.0, 1.0] range. Once δ and ν have been calculated, the fitness of morphology m is obtained as follows:

$$fitness_m = \frac{1}{2}\delta_m + \frac{1}{2}\nu_m \tag{5}$$

Figure 6 illustrates violin plots comparing the fitness of morphologies generated by AFPO, NEAT and HyperNEAT under 500 phase offsets (generated a priori at random) for the new fitness function. Each violin shows the minimum, maximum, median and kernel density estimation of the frequency distribution of values across 500 runs (i.e., phase offset scenarios).

When AFPO is applied (see Fig. 6a), all morphologies demonstrate significant differences (Shapiro-Wilk test, Kruskal-Wallis test; $p < 0.01$). Consequently, it is feasible to rank the performance of the morphologies created by AFPO: Morphology 3 > Morphology 1 > Morphology 4 > Morphology 5 > Morphology 2 (Dunn's test: $p < 0.01$). Furthermore, under NEAT (see Fig. 6b), significant differences can be observed among generated morphologies (Shapiro-Wilk test, Kruskal-Wallis test; $p < 0.01$). Thus, the performance of these morphologies can be ranked as: Morphology 4 > Morphology 3 > Morphology 1 > Morphology 2 > Morphology 5 (Dunn's test: $p < 0.01$). Regarding the morphologies generated by HyperNEAT (see Fig. 6c), significant differences exist among them (Shapiro-Wilk test, Kruskal-Wallis; $p < 0.01$). Hence, a performance rank of the morphologies created by HyperNEAT can be stated: Morphology 5 > Morphology 1 > Morphology 2 > Morphology 3 > Morphology 4 (Dunn's test: $p < 0.01$).

Figure 7 depicts violin plots comparing the performance of the fittest morphology generated by AFPO, NEAT and HyperNEAT. The Kruskal-Wallis test allows the suggestion of significant differences among the fittest morphologies ($p < 0.01$). Hence, the performance can be ranked as follows: HyperNEAT > NEAT > AFPO (Dunn's test: $p < 0.01$).

The narrow data distribution observed in the NE produced morphologies suggests that their responses are similar, regardless of the controller scenario. This can be interpreted as a capacity to perform similarly, despite the synchronisation of the controller. On the other hand, the morphology created by AFPO has a more wide data distribution, suggesting the morphology has a suitable response in only specific scenarios, where the controller is implemented exactly as designed. Consequently, the robustness of this morphology is not comparable to the ones demonstrated by the morphologies produced by NEAT and HyperNEAT.

Results indicate that despite the lack of geometrical aspects of the problem domain that could be embodied in the design of the substrate, the morphologies generated by HyperNEAT have a better trade-off between displacement and the number of voxels than those produced by AFPO and NEAT. This can be justified by the fact that HyperNEAT can produce minimal (i.e., lower volume) and robust morphologies with suitable displacement capabilities under several different controller scenarios. In contrast, AFPO and NEAT create morphologies that in some controller scenarios

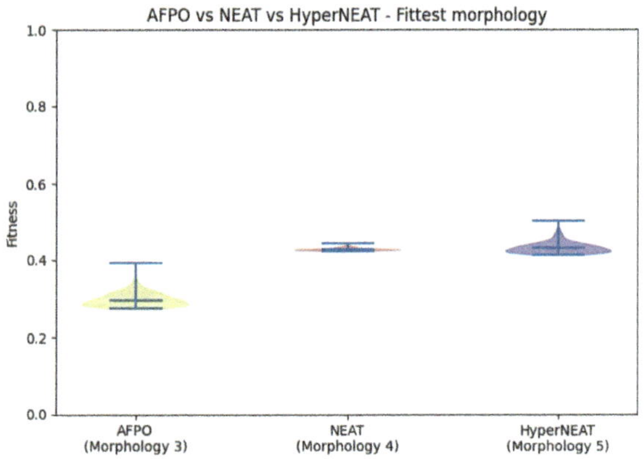

Fig. 7 Fittest morphology under: AFPO, NEAT, and HyperNEAT

Table 3 Number of voxels, mean displacement and fitness value of the fittest morphologies produced by AFPO, NEAT, and HyperNEAT

Approach	Number of voxels	Mean displacement	Fitness value
AFPO	224	2.837	0.297
NEAT	128	4.298	0.429
HyperNEAT	116	3.686	0.432

reach a suitable displacement, however, the number of voxels is higher (i.e., higher volume) than that of the HyperNEAT morphologies (see Table 3).

Finally, Fig. 8 depicts the fittest morphologies generated by all approaches, with red cubes representing active voxels and blue representing passive ones. Under AFPO (Fig. 8a), the morphology contains no passive voxels, and its shape is curve-like (i.e., there are no gaps within the morphology). Furthermore, NEAT produced a smaller and thinner morphology (Fig. 8b), which is a feature that arguably allows it to have a consistent displacement regardless of the phase offset scenario. Despite the morphology being located on the top of the available design area, it is noteworthy that the final displacement is measured from the position that the morphology has after one initial second, allowed for morphologies to settle under gravity (enough to reach the simulated floor) and a further ten simulated seconds (resulting to a simulation of a total of eleven seconds). Regarding HyperNEAT, the morphology generated

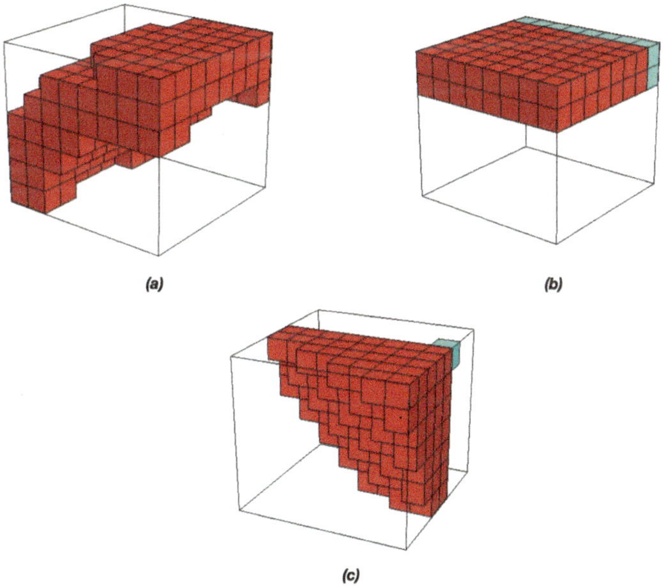

Fig. 8 Fittest morphology generated by: **a** AFPO; **b** NEAT; and **c** HyperNEAT

(Fig. 8c) is a solid triangular-like prism with one passive voxel at the top. Arguably, the geometry (i.e., the voxel asymmetry towards the top vertex) exhibited by this morphology produces stability during displacement, regardless of the phase offset scenario.

Although all approaches employ the same hardware, the computational time differs. Under AFPO, the mean time spent for each evolutionary run is 14.70 h, with a standard deviation of 0.15. Furthermore, when NEAT is utilised, the mean time spent for each evolutionary run is 4.74 h, with a standard deviation of 0.18. Finally, under HyperNEAT, the mean time spent is 6.82 h with a standard deviation of 0.15 for each evolutionary run. Despite the different execution times (due to increased functionalities of the AFPO baseline code, like saving checkpoints), the robustness of the generated morphologies is the focus of this work. Further investigation is required to compare the efficiency in terms of execution times for the code implementing the alternative methodologies. Arguably, the software architecture implemented for NEAT and HyperNEAT can utilise the hardware more efficiently, since the client-server model is adopted, resulting to significantly reduced computation time during evolution.

5 Conclusions

The work presented here explored the suitability of NEAT and HyperNEAT in generating suitable BHM morphologies for maximising displacement in a physics simulator. The capabilities of NEAT and HyperNEAT were compared against the AFPO approach, a well-known and tested multi-objective optimisation approach. The fitness functions considered were (i) the maximum displacement and (ii) the trade-off between the number of voxels composing morphologies along with their displacement. Under 500 different scenarios with different predefined phase offsets of the BHM controllers, five morphologies generated by each approach were compared. Results indicate that, generally, the morphologies generated by HyperNEAT require fewer voxels to reach similar or farther displacements than morphologies produced by AFPO and NEAT. Moreover, compact morphologies tend to show a consistent displacement regardless of the phase offset controller scenario; this can be interpreted as those morphologies being more robust than bulky structures.

From a biomechanical perspective, these results match one of the most common mechanisms of locomotion observed in soft-bodied animals (e.g., earthworms, sea cucumbers, caterpillars, and snails): *peristalsis*, which consists of alternating waves of muscle contraction and relaxation move along the body, anchoring the body to the substratum only in widened regions [32].

Future work directions can be outlined based on the results and insights obtained by this research. For instance, one aspect of future work can be adding more periodic activation functions (e.g., *cosine*, *tangent*) to explore additional morphological patterns produced by CPPNs. Furthermore, fine-tuning the GA may improve the performance. For example, increasing the number of individuals and number of generations may be advantageous, whereas minimal initialisation of CPPNs have been proved to accelerate the computations and not affect the efficiency of morphologies [29]. Another relevant aspect to consider is the design of the substrate; arguably, a broader exploration of the number of layers, neurons and activation functions may lead to better performance of HyperNEAT. Moreover, an approach that can evolve the geometry of the substrate (i.e., evolve the location of every neuron and the pattern of weights among them) can be implemented to improve the performance of morphologies; this approach is known as *ES-HyperNEAT* [17].

In conclusion, given the intricacies involved in the design process of BHMs, it is imperative to subject HyperNEAT methodology to further testing on more intricate challenges. This should encompass not only the simulation of the actuator itself, but, also the interactions with the passive components of a desired BHM system, i.e., in applications of medical devices.

Acknowledgements This project has received funding from the European Union's Horizon Europe research and innovation programme under grant agreement No. 101070328. UWE researchers were funded by the UK Research and Innovation grant No. 10044516. **Disclosure of Interests.** The authors have no competing interests to declare that are relevant to the content of this article.

References

1. Auerbach, J.E., Bongard, J.C.: Evolving complete robots with cppn-neat: The utility of recurrent connections. In: Proceedings of the 13th Annual Conference on Genetic and Evolutionary Computation. GECCO '11, Association for Computing Machinery (2011). https://doi.org/10.1145/2001576.2001775
2. Bartlett, N.W., Tolley, M.T., Overvelde, J.T., Weaver, J.C., Mosadegh, B., Bertoldi, K., Whitesides, G.M., Wood, R.J.: A 3d-printed, functionally graded soft robot powered by combustion. Science **349**(6244), 161–165 (2015)
3. Evjemo, L.D., Gjerstad, T., Grøtli, E.I., Sziebig, G.: Trends in smart manufacturing: Role of humans and industrial robots in smart factories. Current Robotics Reports **1**(2), 35–41 (2020). https://doi.org/10.1007/s43154-020-00006-5
4. Hawkes, E.W., Blumenschein, L.H., Greer, J.D., Okamura, A.M.: A soft robot that navigates its environment through growth. Science Robotics **2**(8), eaan3028 (2017)
5. Hiller, J., Lipson, H.: Dynamic simulation of soft multimaterial 3d-printed objects. Soft robotics **1**(1), 88–101 (2014)
6. Kimura, T., Jin, Z., Niiyama, R., Kuniyoshi, Y.: Evolving soft robots to execute multiple tasks with combined-cppn-neat. In: 2016 IEEE/SICE International Symposium on System Integration (SII). pp. 409–414 (2016). https://doi.org/10.1109/SII.2016.7844033
7. Kriegman, S., Blackiston, D., Levin, M., Bongard, J.: A scalable pipeline for designing reconfigurable organisms. Proceedings of the National Academy of Sciences **117**(4), 1853–1859 (2020). https://doi.org/10.1073/pnas.1910837117
8. Laschi, C., Cianchetti, M., Mazzolai, B., Margheri, L., Follador, M., Dario, P.: Soft robot arm inspired by the octopus. Advanced robotics **26**(7), 709–727 (2012)
9. Lee, C., Kim, M., Kim, Y.J., Hong, N., Ryu, S., Kim, H.J., Kim, S.: Soft robot review. International Journal of Control, Automation and Systems **15**, 3–15 (2017)
10. Lipson, H.: Challenges and opportunities for design, simulation, and fabrication of soft robots. Soft Robotics **1**(1), 21–27 (2014)

11. Mazzolai, B., Meloni, G., Degl'Innocenti, A.: Can a robot grow? plants give us the answer. In: Bioinspiration, Biomimetics, and Bioreplication 2017. vol. 10162, pp. 24–33. SPIE (2017)
12. Mestre, R., Patiño, T., Sánchez, S.: Biohybrid robotics: From the nanoscale to the macroscale. Wiley Interdisciplinary Reviews: Nanomedicine and Nanobiotechnology **13**(5), e1703 (2021)
13. Morimoto, Y., Onoe, H., Takeuchi, S.: Biohybrid robot powered by an antagonistic pair of skeletal muscle tissues. Science robotics **3**(18), eaat4440 (2018)
14. Nidorf, D.G., Barone, L., French, T.: A comparative study of neat and xcs in robocode. In: IEEE Congress on Evolutionary Computation. pp. 1–8 (2010). https://doi.org/10.1109/CEC.2010.5586087
15. Perez, S.A.H., Harada, K., Mitsuishi, M.: Haptic virtual fixtures to assist endonasal micro robotic surgery through virtual reality simulation. In: 2018 International Symposium on Micro-NanoMechatronics and Human Science (MHS). pp. 1–3 (2018). https://doi.org/10.1109/MHS.2018.8886945
16. Reyes, P., Escobar, M.J.: Neuroevolutive algorithms for learning gaits in legged robots. IEEE Access **7**, 142406–142420 (2019). https://doi.org/10.1109/ACCESS.2019.2944545
17. Risi, S., Stanley, K.O.: An Enhanced Hypercube-Based Encoding for Evolving the Placement, Density, and Connectivity of Neurons. Artificial Life **18**(4), 331–363 (10 2012), https://doi.org/10.1162/ARTL_a_00071
18. Rose, D.C., Lyon, J., de Boon, A., Hanheide, M., Pearson, S.: Responsible development of autonomous robotics in agriculture. Nature Food **2**(5), 306–309 (2021). https://doi.org/10.1038/s43016-021-00287-9
19. Ruebsamen, G.D.: Evolving intelligent embodied agents within a physically accurate environment. California State University, Long Beach (2002)
20. Rus, D., Tolley, M.T.: Design, fabrication and control of soft robots. Nature **521**(7553), 467–475 (2015)
21. Sadeghi, A., Mondini, A., Del Dottore, E., Mattoli, V., Beccai, L., Taccola, S., Lucarotti, C., Totaro, M., Mazzolai, B.: A plant-inspired robot with soft differential bending capabilities. Bioinspiration & biomimetics **12**(1), 015001 (2016)
22. Schmidt, M.D., Lipson, H.: Age-fitness pareto optimization. In: Proceedings of the 12th Annual Conference on Genetic and Evolutionary Computation. pp. 543–544. Association for Computing Machinery, New York, NY, USA (2010). https://doi.org/10.1145/1830483.1830584
23. Schulz, A., Sung, C., Spielberg, A., Zhao, W., Cheng, R., Grinspun, E., Rus, D., Matusik, W.: Interactive robogami: An end-to-end system for design of robots with ground locomotion. The International Journal of Robotics Research **36**(10), 1131–1147 (2016). https://doi.org/10.1177/0278364917723465
24. Song, Y.L., Chiu, W.Y., Manoharan, S.H.: Mixed compositional pattern-producing network-neuroevolution of augmenting topologies method for the locomotion control of a snake-like modular robot. In: 2023 SICE International Symposium on Control Systems (SICE ISCS). pp. 93–100 (2023). https://doi.org/10.23919/SICEISCS57194.2023.10079203
25. Stanley, K.O., D'Ambrosio, D.B., Gauci, J.: A Hypercube-Based Encoding for Evolving Large-Scale Neural Networks. Artificial Life **15**(2), 185–212 (04 2009). https://doi.org/10.1162/artl.2009.15.2.15202
26. Stanley, K.O., Miikkulainen, R.: Evolving neural networks through augmenting topologies. Evolutionary Computation **10**(2), 99–127 (2002). https://doi.org/10.1162/106365602320169811
27. Tibermacine, A., Djedi, N.: Neat neural networks to control and simulate virtual creature's locomotion. In: 2014 International Conference on Multimedia Computing and Systems (ICMCS). pp. 9–14 (2014). https://doi.org/10.1109/ICMCS.2014.6911392
28. Trivedi, D., Rahn, C.D., Kier, W.M., Walker, I.D.: Soft robotics: Biological inspiration, state of the art, and future research. Applied bionics and biomechanics **5**(3), 99–117 (2008)
29. Tsompanas, M.A.: Incremental growth on compositional pattern producing networks based optimization of biohybrid actuators. In: International Conference on the Applications of Evolutionary Computation (Part of EvoStar). pp. 275–289. Springer (2024)

30. Tsompanas, M.A., Balaz, I.: Outline of an evolutionary morphology generator towards the modular design of a biohybrid catheter. Frontiers in Robotics and AI **11** (2024)
31. Wang, H., Totaro, M., Beccai, L.: Toward perceptive soft robots: Progress and challenges. Advanced Science **5**(9), 1800541 (2018)
32. Yekutieli, Y., Flash, T., Hochner, B.: Biomechanics: Hydroskeletal. In: Squire, L.R. (ed.) Encyclopedia of Neuroscience, pp. 189–200. Academic Press, Oxford (2009). https://doi.org/10.1016/B978-008045046-9.01948-3

The Holocron: Leveraging a Dynamic 3D Matrix for Efficient Data Storage and Retrieval in Strategic Management

Matthew Visconti and Stelios Kapetanakis

Abstract The Holocron 3D matrix serves as a pivotal component in data storage and retrieval, designed to manage and streamline strategic data with exceptional efficiency. This dynamic data structure supports the continuous evolution of business strategies by accommodating real-time updates and seamlessly integrating new strategic connections identified by the Graph Neural Network (GNN). The matrix's architecture allows for the organization of complex, multi-level strategic threads, facilitating swift and accurate data access. By ensuring the integrity and coherence of strategic information across the Corporate, Business, and Functional levels, the Holocron 3D matrix enhances decision-making processes. This paper explores the underlying mechanisms of the Holocron, highlighting its role in maintaining a robust, adaptive framework that aligns with the overarching objectives of strategic management within distributed systems. Through this dynamic integration, the Holocron 3D matrix not only bridges the gap between static strategic planning and adaptive strategic thinking but also drives organizations toward achieving strategic excellence and innovation.

Keywords Strategic management · Dynamic integration · Data storage

1 Introduction

In today's dynamic business environment, efficient data storage and retrieval are paramount to effective strategic management. Organizations must navigate complex data landscapes to make informed decisions that align with their long-term objectives while adapting to immediate challenges [1]. The ability to manage and

M. Visconti (✉) · S. Kapetanakis
Middlesex University, London, England
e-mail: matthewvisconti@outlook.com

Distributed Analytics Solutions, London, UK

S. Kapetanakis
e-mail: stelios@distributedanalytics.co.uk

N. Polatidis et al. (eds.), *Intelligent Distributed Computing XVII*, Studies in Computational Intelligence 1203, https://doi.org/10.1007/978-3-031-87639-4_2

access strategic data efficiently is crucial for maintaining competitive advantage and achieving organizational goals.

This paper introduces the Holocron 3D matrix, a novel data structure designed to enhance the management of strategic information across multi-level business frameworks. The Holocron 3D matrix acts as a dynamic repository, capable of storing, organizing, and retrieving vast amounts of strategic data in real-time [2]. By leveraging this advanced data structure, organizations can ensure that their strategic decisions are informed by the most current and relevant information, thereby improving coherence and synergy across different organizational levels.

The primary objective of this paper is to explore the design and functionality of the Holocron 3D matrix and demonstrate its efficacy in managing strategic data. We will discuss how the matrix integrates with a Graph Neural Network (GNN) to identify and establish novel strategic connections, thereby facilitating adaptive and dynamic strategic planning. Through this examination, we aim to highlight the significance of the Holocron 3D matrix in revolutionizing data management practices within distributed systems, ultimately driving organizations toward strategic excellence and innovation.

2 The Rationale Behind Using a 3D Matrix for Strategic Data Management

The Holocron 3D matrix offers a novel approach to strategic data management, particularly suited for distributed systems. The rationale behind employing a 3D matrix structure lies in its ability to organize and manage multi-dimensional data efficiently. In the context of strategic management, organizations operate across multiple levels; Corporate, Business, and Functional, each with distinct data requirements and interdependencies. A 3D matrix provides a natural and intuitive way to map these levels, capturing the complexity and nuances of strategic interactions [3].

By leveraging the spatial properties of a 3D matrix, the Holocron can store data points representing various strategic entities and their interconnections across different organizational levels. This structure allows for dynamic updates and real-time data retrieval, ensuring that strategic decisions are based on the most current and relevant information. Furthermore, the 3D matrix facilitates the integration of a Graph Neural Network (GNN), which can analyse and identify novel strategic connections, enhancing the organization's ability to adapt to new challenges and opportunities.

The use of a 3D matrix also supports scalability, as it can accommodate increasing volumes of data and more complex strategic relationships without significant degradation in performance [4]. This scalability is crucial for distributed systems, where the data landscape continually evolves. Additionally, the inherent flexibility of the 3D matrix allows for seamless integration of new data points and strategic insights, maintaining the coherence and alignment of organizational goals across different levels [5].

In summary, the Holocron 3D matrix represents an innovative data storage solution that addresses the specific needs of strategic data management in distributed systems. Its ability to efficiently organize, update, and retrieve complex multi-dimensional data makes it an invaluable tool for organizations striving to achieve strategic excellence and innovation (Fig. 1).

Holocron Data Structure

In a structured organisation, the three organisation levels: Corporate, Business, and Functional, each possess distinct roles and characteristics. The Corporate Level, akin to the "Brain" of the organisation, focuses on long-term strategic goals and overall direction. The Business level, acting as the "heart", dynamically engages with the market, driving operational strategies that respond to external factors. Finally, the Functional level, serving as the "limbs", handles the day-to-day operations and tasks necessary to support the business activities [6] (Fig. 2).

These inherent characteristics drive each level to gravitate towards specific categories of strategies. The Corporate level typically aligns with Analytical/Systematic strategies, such as those from the Planning and Design Schools, which involve logical, systematic planning. The Business level tends to favour Environmentally influenced strategies, including those form the Entrepreneurial and Learning Schools, which adapt to market dynamics. Meanwhile, the Functional Level focuses on Internal

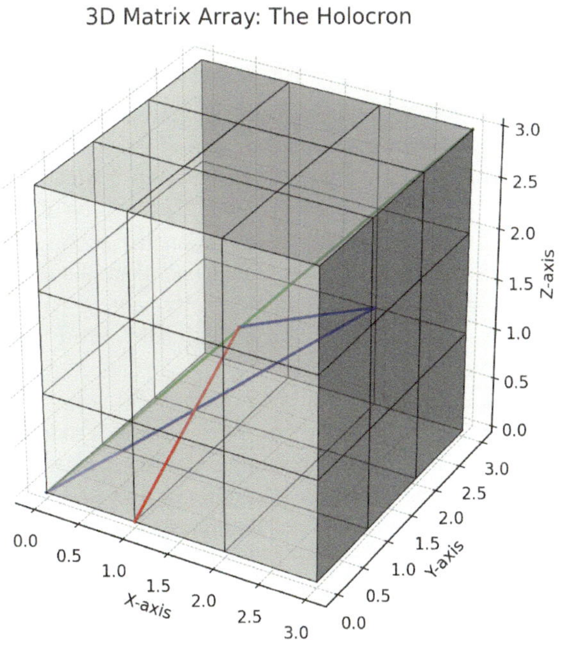

Fig. 1 Diagram of the 3D Matrix Array with embedded threads and varying depths

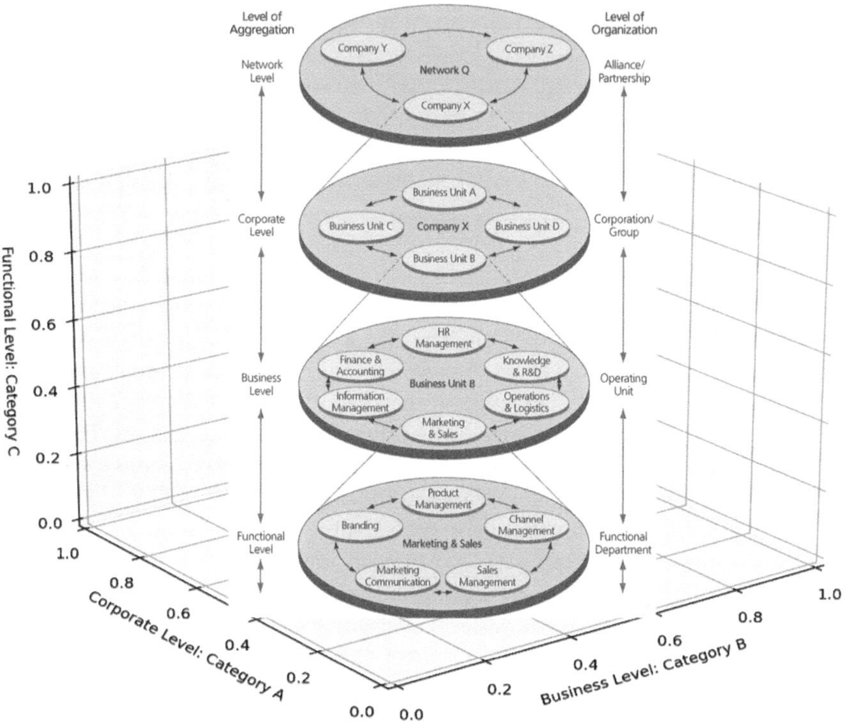

Fig. 2 Multi-levelled organizational diagram superimposed on the levels represented as XYZ coordinate axis

strategies, like those from the Cognitive and Culture Schools, optimising internal processes and capabilities.

Within these categories, strategies are evaluated and placed on a rationality spectrum, where 100 represents the highest degree of rationality. Each category encompasses strategies with varying rationality scores. To create effective strategic links between different business level, one must identify and align strategies across these categories based on their rationality scores. This process involves examining the relative closeness of strategies' rationality scores and determining their potential for integration.

For example, a "Market Expansion Strategy" at the corporate level, with a rationality score of 90, can be linked with a "Product Diversification" strategy at the Business level, which has a rationality score of 75. This alignment is achieved by traversing the Holocron 3D matrix to find strategies at the Business and Functional levels that meet the defined criteria, such as having a rationality score above a certain threshold and belonging to a relevant category. In this case, the "Operational Efficiency Improvement" strategy at the Functional Level, with a score of 65, does not meet the threshold, and thus, is not considered for this link.

By establishing these connections, organisations can align strategies across different levels, ensuring that each level's strategic approach supports and enhances the others. This method facilitates coherent and integrated strategic planning, leveraging the strengths and addressing the unique challenges at each organisational level. The Holocron 3D matrix serves as a powerful tool to manage, retrieve and analyse these strategic links, enabling organisations to adapt to evolving business environments and make informed, strategic decisions.

The Holocron 3D matrix is designed to manage and retrieve strategic data efficiently, accommodating the fluid nature of business strategies that continuously evolve due to shifting market dynamics, organisational changes, and competitive landscapes.

Structure of the Holocron 3D Matrix

The 3D matrix encapsulates data points representing various strategic entities across three dimensions, corresponding to different levels of the organisational hierarchy: Corporate, Business, and Functional.

Dimensions:

- X-axis: Represent the Corporate level strategies.
- Y-axis: Represents the Business level strategies.
- Z-axis: Represents the Functional level strategies.

How Data is Organised, Stored and Retrieved.

Data Organisation:

- Nodes: Each cell (node) in the 3D matrix contains a data point representing a strategic element or a set of elements.
- Edges: Connections between nodes represent relationships or threads of congruence between different strategic elements.

Data Storage:

- Attributes: Each node stores attributes relevant to its strategic element, such as its rationality score, category, and associated metrics.
- Metadata: Additional metadata includes timestamps, source references, and update logs to track changes over time [7].

Data Retrieval

Indexing: The matrix is indexed based on the three dimensions, allowing for quick access to data points using their coordinates.

Query Mechanism: A query mechanism enables the retrieval of data points or threads based on specific criteria, such as a particular strategy's rationality score or category [8].

Corporate Level (X-axis):

- Node (1, 0, 0): Represents high level rational long-term strategic thinking such as "Market Expansion Strategy."
- Attributes: Rationality Score: 90, Category: Analytical.

Business Level (Y-axis):

- Node (0, 1, 0): Represents a business-level strategy, such as "Product Diversification."
- Attributes: Rationality Score: 75, Categoriy: Environmental.

Functional Level (Z-axis):

- Node (0, 0, 1): Represents a functional-level strategy, such as "Operational Efficiency Improvement."
- Attributes: Rationality Score: 65, Category: Internal.

In developing a strategic framework for organisational management each organisational level, Corporate, Business and Functinoal has its distinct characteristics and strategic focuses. These levels can be mapped to different categories of strategies, which are then evaluated and scored on a rationality spectrum.

These scores place each strategy on a rationality spectrum, with 100 representing the highest degree of rationality. The strategies are then categorised and aligned to the corresponding organisational levels:

(1) Corporate Level (X-axis): Analytical/Systematic strategies.
(2) Business Level (Y-axis): Environmentally influenced strategies.
(3) Functional Level (Z-axis): Internal Focus strategies.

The Holocron 3D matrix captures this structure by mapping each strategy to a specific coordinate on a 3D plane, where each axis represents a different organisational level. This setup allows for an accurate representation of how strategies align across the corporate, business and functional levels.

To create and access threads (links) between these different levels and strategy categories, various methodologies can be employed, each mapped to an appropriate traversal algorithm [9]:

Categorising Regions and Mapping Traversal Algorithms in the Holocron 3D Matrix

To optimise the retrieval of strategic threads in the Holocron 3D matrix, the matrix is categorised into general regions based on positions, depths, and locations. Each region is mapped to appropriate traversal algorithms, ensuring efficient access to threads.

Categorising Regions

The 3D matrix is divided into distinct regions, each with specific characteristics that influence the choice of traversal algorithm. These regions can be broadly categorised as:

(1) **Shallow Regions**:

- These are near the surface or top layers of the matrix.
- Strategies immediately related to high-level corporate strategies.
- Breadth-First Search (BFS) is used for quickly finding the nearest relevant nodes [10].

(2) **Intermediate Regions**:

- These are mid-level layers that bridge shallow and deep regions.
- Strategies that connect business-level strategies to functional-level strategies.
- Dijkstra's Algorithm is used for finding optimal paths in weighted graphs [11].

(3) **Deep Regions**:

- Located deeper within the matrix, involving more complex interconnections.
- Deeply nested strategies within the functional level.
- Depth-First Search (DFS) is employed for a thorough exploration of all possible paths [12].

3 Leveraging the Holocron 3D Matrix in Distributed Systems

Distributed systems, composed of multiple interconnected nodes, are crucial for processing and analysing large datasets in real-time. These systems are designed to manage complex data structures and enable dynamic decision-making across modern organisations [13]. The Holocron 3D matrix integrates seamlessly into this framework, mapping each organisational level to specific nodes within the distributed system. This integration ensures efficient storage, management, and processing of data relevant to each level, enhancing the overall functionality of the distributed system.

In this setup, the Corporate level is represented along the X-axis, serving as the central node that sets overarching long-term goals. The Business level, mapped to the Y-axis, acts as intermediary nodes that dynamically engage with the market environment. The Functional level, corresponding to the Z-axis, functions as peripheral nodes that handle day-to-day operations. This clear mapping ensures that strategic data is organised and accessible, allowing each level to perform its role effectively.

To create and access strategic threads within the Holocron 3D matrix, various traversal algorithms are employed. These methodologies include clustering and aggregation using Breadth-First Search (BFS), direct alignment based on rationality scores using Depth-First-Search (DFS), and bridging gaps between disparate strategies using Dijkstra's Algorithm [10–12]. Each method offers a unique approach to linking strategies across different levels. Ensuring that the organisations strategies are coherent and aligned with the corporate goals.

For instance, clustering and aggregation using BFS groups similar strategies within each category based on rationality scores, linking nodes that fall into the same or adjacent clusters. Direct alignment using DFS connects strategies from different levels with closely aligned rationality scores, ensuring direct and logical connections. Dijkstra's Algorithms is employed to bridge the gaps between strategies with disparate scores, creating a cohesive path through intermediate nodes.

By levering these traversal algorithms, the Holocron 3D matrix efficiently manages the nuanced complexity of strategies at different organisational levels. This approach facilitates dynamic and adaptive planning, allowing the organisations to quickly respond to changing market conditions and internal shifts. It also supports efficient data management, enhancing overall performance and scalability of the distributed system.

In practical application, consider a "Market Expansion Strategy" at the corporate level with a high rationality score. By defining criteria such as a rationality score threshold and relevant strategy categories, the Holocron 3D matrix can traverse the matrix to identify and link related strategies at the Business and Functional levels. This process results in a cohesive strategic thread that integrates insights from multiple levels, ensuring alignment with the overarching goals and facilitating informed decision-making.

Dynamic Interaction and Updating

The Holocron 3D matrix is designed to manage and retrieve strategic data efficiently, making it well-suited for dynamic business environments. To ensure that the matrix can handle real-time data integration and provide up-to-date strategic insights, several advanced techniques and tools are employed.

Dynamic Data Structures

To accommodate the fluid nature of business strategies, the Holocron utilises dynamic data structures. Sparse matrices are particularly useful, as they store only non-zero elements, optimising storage for large 3d matrices with many empty cells. Tools like 'SciPy' in Python ('scipy.sparse' module) facilitate this efficient storage [14]. Additionally, dynamic arrays from the 'NumPy' library automatically resize to accommodate new elements, ensuring the 3D matrix can grow as new data points are added without significant overhead [15].

Real-Time Data Integration

Real-time data integration is crucial for continuously updating the Holocron with the latest information. Stream processing allows for the continuous input, processing, and analysis of data. Tools such as Apache Kafka, Apache Flink, or Apache Storm enable real-time data streams to be managed effectively [16]. For integrating large volumes of data at regular intervals, batch processing tools like Apache Spark or Hadoop are used [17]. These tools ensure that the Holcron remains current, reflecting the latest strategic configurations and enabling swift identification of novel alignments.

Database Systems

The Holocron leverages NoSQL databases to handle large volumes of unstructured data and support dynamic schemas. MongoDB, for example, is used for document-oriented storage, while Cassandra offers wide-column storage capabilities. Additionally, graph database like Neo4j are employed to manage and query the interconnected data points within the 3D matrix [18, 19]. These databases are designed to handle complex relationships and provide efficient querying mechanisms, which are essential for dynamic strategic management.

4 Implementation and Integration

Integrating New Threads into the Holocron 3D Matrix

Integrating new threads into the Holocron 3D matrix involves a systematic approach to ensure that new connections seamlessly align with pre-existing ones, thereby maintaining the coherence of the strategic framework. The process begins with identifying the nodes that will be part of the new thread. These nodes could either be existing within the matrix or may need to be added as new entries. It is crucial to update the attributes of these nodes to reflect and most current data, which may include rationality scores or other relevant metrics. The integration of new connections involves updating the adjacency information within the matrix or the graph structure, which may entail adding new edges or modifying existing ones to incorporate the new strategic relationships.

Practical Implementation

To practically implement this process, the following steps are followed: identifying nodes and edges, updating node attributes and superimposing connections. First, the starting and ending nodes of the new connections are determined. Ensuring these nodes exist within the matrix is essential, and new nodes are added if they are not already present. The attributes of these nodes are then updated to incorporate the latest data, ensuring that the strategic insights remain relevant and accurate. Finally, new connections are integrated into the matrix by adding the new edges between identified nodes, ensuring they do not conflict with existing ones. This process maintains the dynamic and adaptive nature of the Holocron 3D matrix.

Detailed Example

Consider an example where a starting strategy at the Corporate level, such as "Market Expansion Strategy" (Node (1,0,0)), is identified. The GNN might identify new strategies, such as a "Global Partnership Initiative" (Node (2,1,1)). The criteria for establishing new connections include a rationality score greater than 70 and a category aligned with Analytical or Environmental strategies. Using traversal algorithms, nodes and edges are identified, and attributes are updated as necessary. The new connections are then superimposed onto the matrix by adding edges between the identified nodes, integrating them seamlessly into the existing structure [7].

Python Implementation

This process can be implemented in Python using a combination of scipy.sparse for the matrix and neo4j for graph connections. The Holocron 3D matrix is initialized as a sparse matrix, which allows for efficient storage and retrieval of data. A function is defined to update the matrix with new node attributes, and another function to add relationships in the Neo4j graph database. This ensures that both the matrix and the graph are kept in sync, reflecting the latest strategic connections.

Superimposing Connections

To integrate these new connections effectively, the rationality scores and categories of the nodes are updated. For instance, if the starting node is "Market Expansion Strategy" with a rationality score of 90, and the new node is "Global Partnership Initiative" with a score of 85, these attributes are reflected in the Holocron matrix. Connections are then superimposed using Neo4j to ensure seamless integration without conflicts. This method maintains the dynamic adaptability of the Holocron 3D matrix, allowing for continuous updates and real-time strategic alignment.

5 Integrating Novel Connections Into the Holocron 3D Matrix

Conceptual Overview

In the context of integrating the Holocron 3D matrix with a Graph Neural Network (GNN), updates involve several key aspects to ensure that the matrix remains current and reflective of the latest strategic insights. These updates typically include the insertion of new nodes and edges, updating node attributes, and creating or managing connections (threads) within the matrix. This process ensures that both new strategies and refined relationships between existing strategies are accurately captured.

Inserting New Nodes and Edges

While the GNN primarily identifies novel connections between pre-existing strategies, there are instances where entirely new strategies are discovered and need to be integrated into the Holocron. This involves inserting new nodes into the 3D matrix, which represents these newly identified strategies. Additionally, new edges are created to represent the relationships between these new nodes and the existing ones. For example, if the GNN identifies a "Global Partnership Initiative" as a new strategy, this node would be added to the matrix, enhancing the strategic framework.

Updating Attributes

Even if the nodes themselves are pre-existing, the attributes of these nodes may require updates based on the latest insights provided by the GNN. For instance, the GNN might refine the rationality scores or update the categories based on new data. These updated attributes ensure that the nodes accurately reflect the most current strategic information. For example, if the rationality score of the "Market Expansion Strategy" is refined from 90 to 95, this updated score is crucial for maintaining the accuracy of the matrix.

Creating and Managing Connections (Threads)

The primary role of the GNN is to identify and validate novel connections (threads) between existing strategies. These connections must be accurately recorded in the Holocron. This involves updating the adjacency information or the graph structure within the matrix to reflect these new edges. For example, if the GNN identifies a novel connection between "Market Expansion Strategy" and "Product Diversification," this connection is added to the Holocron, ensuring that the strategic relationships are updated.

Detailed Example of Updates

Consider an example where the GNN identifies a novel thread connecting the "Market Expansion Strategy" (Node (1,0,0)) with "Product Diversification" (Node (0,1,0)). Additionally, the GNN refines the rationality score of the "Market Expansion Strategy" from 90 to 95. In this case, the following steps are undertaken:

1. **Inserting New Nodes (if applicable)**: if the GNN identifies a new strategy, it is added to the Holocron:

```
holocron[2, 0, 0] = 0.85
```

2. **Updating Attributes**: the rationality score of the existing node is updated:

```
2 usages
def update_holocron(matrix, coordinates, value):
    matrix[coordinates] = value
```

3. **Creating Connections**: the new connection identified by the gnn is added to the holocron

```
def add_relationship(tx, start_node, end_node):
    tx.run("MERGE (a:Node {x: $x1, y: $y1, z: $z1}) "
           "MERGE (b:Node {x: $x2, y: $y2, z: $z2}) "
           "MERGE (a)-[r:CONNECTED_TO]->(b)",
           x1=start_node[0], y1=start_node[1], z1=start_node[2],
           x2=end_node[0], y2=end_node[1], z2=end_node[2])
```

The Holocron 3D matrix remains dynamically updated, reflecting both new strategies and refined attributes, as well as new connections identified by the GNN. This ensures that the matrix accurately represents the latest strategic insights and relationships, enabling informed and adaptive decision-making within the organization. This approach highlights the dynamic nature of the Holocron, allowing it to evolve continuously and maintain relevance in a rapidly changing business environment.

Identifying and Accessing Specific Threads in the Holocron 3D Matrix

In the context of the Holocron 3D matrix, threads (or connections between nodes) can be accessed through unique identifiers or coordinates. This structured approach ensures efficient management and retrieval of strategic data.

Using Unique Identifiers

Each thread within the Holocron matrix can be assigned a unique identifier (ID) that encapsulates the connection between specific nodes. This unique ID combines the coordinates of the connected nodes, allowing for quick referencing and access. For instance, a unique ID can be generated using a function that combines the start and end node coordinates, resulting in a format like "1, 0, 0, 2, 1, 1". These ID's can be stored in a database along with detailed attributes of the connection, such as weight, timestamp, and other relevant properties.

Example Implementation

Consider a scenario where a thread connects a "Market Expansion Strategy" at node (1, 0, 0) with "Product Diversification" at node (2, 1, 1). A unique ID is generated and stored in a dictionary or a database. This storage approach ensures that each thread's attributes are easily accessible using its unique ID.

Using Coordinates for Traversal

Traversal algorithms are employed to navigate the matrix using the coordinates of the nodes. This involves identifying the start and end coordinates and utilsing specific algorithms such as Breadth-First Search (BFS) or Depth-First Search (DFS) to traverse the path. For instance, a traversal function can be implemented to move from the start node to the end node, recoding the path taken.

Storing and Retrieving Threads in a Database

To manage the threads efficiently, they can be stored in a database with their unique IDs and associated attributes. Using a database like MongoDB, threads can be inserted and queried efficiently. For example, after generating a unique ID for a thread, it can be stored in the database with its start and end nodes and additional properties. This setup allows for easy retrieval of threads by querying the database with their unique IDs.

Consider a thread with a start node at (1, 0, 0) and an end node at (2, 1, 1). This unique ID "1_0_0_2_1_1" is generated and used to store and thread in the database. To retrieve this thread, a simple query using the unique ID can fetch all the relevant details, ensuring efficient data management.

Dynamic Interaction and Updating

The Holocron 3D matrix supports dynamic interactions and continuous updates. As new data and insgihts are generated from the Graph Neural Network (GNN), the matrix can be updated to reflect these changes. This dynamic capability ensures that the matrix remains current, incorporating new strategic threads and adjusting existing ones as necessary.

Integration with GNN

The GNN outputs novel strategic connections that need to be integrated into the Holocron matrix. This involves mapping the GNN-generated threads to the matrix's existing structure without causing conflicts. By updating the matrix with these new threads, the system remains adaptive and capable of reflecting the latest strategic insights.

Threads are mapped to traversal algorithms based on their general region within the 3D matrix. This mapping ensures that the most efficient algorithm is used for accessing and retrieving threads (Fig. 3).

6 Conclusion

The Holocron 3D matrix represents a significant advancement in data storage and retrieval mechanisms, particularly for strategic management within distributed systems. By leveraging a multi-dimensional matrix structure, the Holocron efficiently organizes and manages complex strategic data across different organizational levels. This innovative approach ensures that data remains accessible and up to date, facilitating real-time decision-making and strategic planning.

The matrix can dynamically expand to accommodate increasing volumes of data, ensuring that the system remains responsive and effective even as the scope and complexity of the data grow. Furthermore, the distributed nature of the Holocron provides redundancy, ensuring that critical data is protected and remains accessible even in the event of hardware failures or other disruptions. This resilience is crucial for

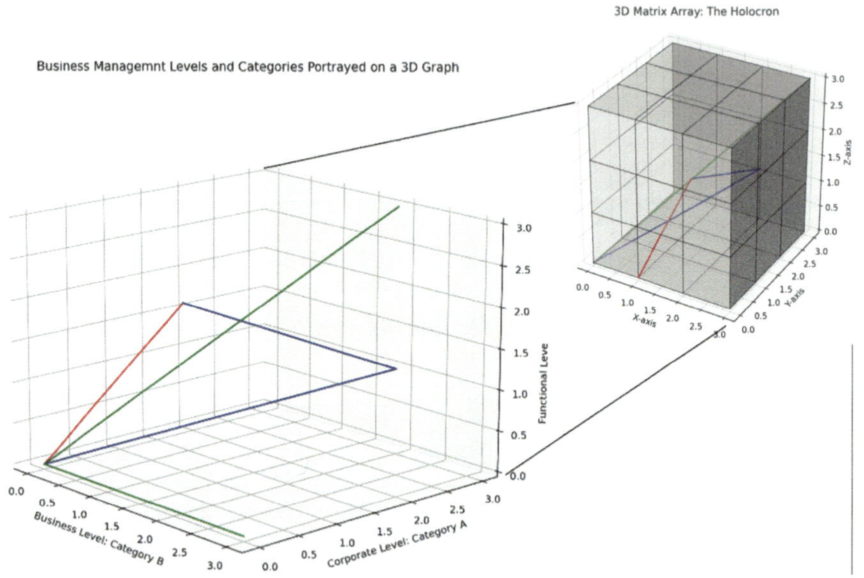

Fig. 3 Accessing threads from Holocron and plotting them on XYZ graph to access concrete strategies

maintaining the integrity and availability of strategic information, thereby supporting continuous and informed decision-making processes.

The integration of the Holocron with Graph Neural Networks (GNNs) enhances its capability to identify novel strategic connections, driving innovation and strategic alignment within organizations. By capturing and analysing the intricate relationships between different strategic entities, the Holocron enables organizations to navigate complex business environments more effectively, ensuring that their strategic objectives are met with precision and coherence.

Future Directions for Enhancing Data Storage and Retrieval Mechanisms

Looking ahead, there are several promising directions for further enhancing the Holocron and similar data storage and retrieval mechanisms. One key area of focus is the integration of advanced machine learning algorithms to further refine the identification and analysis of strategic connections. By incorporating techniques such as reinforcement learning and advanced feature engineering, the system can improve its predictive capabilities and adaptability to changing business environments.

Another potential enhancement is the development of more sophisticated fault tolerance and data replication strategies. This could involve the implementation of distributed ledger technologies, such as blockchain, to ensure even higher levels of data integrity and security [20]. Such advancements would provide additional layers of protection against data corruption and unauthorized access, further enhancing the resilience of the system.

Finally, expanding the applicability of the Holocron to a broader range of industries and use cases can provide valuable insights into its versatility and effectiveness.

Disclosure of Interests. The author has no competing interests to declare. This project was conducted independently without any external funding or affiliations.

References

1. Sathi, V. and Sadhasivan, S., 2020. Accessibility and big data analytics on cloud. *Preprint, available at:* https://www.researchgate.net/publication/347443390_Accessiblity_and_Big_Data_Analytics_on_Cloud.
2. Dichenko, S.A., 2021. An integrity control model for multidimensional data arrays. *Automatic Control and Computer Sciences, 55*(8), pp.1188-1193.
3. Baumann, P., Misev, D., Merticariu, V. and Huu, B.P., 2021. Array databases: Concepts, standards, implementations. *Journal of Big Data, 8*, pp.1-61.
4. Rusu, F., 2023. Multidimensional array data management. *Foundations and Trends® in Databases, 12*(2–3), pp.69–220.
5. Yang, D., Long, C., Xu, H. and Peng, S., 2020, March. A review on scalability of blockchain. In *Proceedings of the 2020 the 2nd International Conference on Blockchain Technology* (pp. 1–6).
6. Meyer, R. and De Wit, B., 1999. Strategy Synthesis: Blending Conflicting Perspectives to Create Competitive Advantage. London: Cengage Learning EMEA
7. Johnson, J., Whittington, R., Regnér, P., Angwin, D., Johnson, G. and Scholes, K., 2020. *Exploring strategy*. Pearson UK.
8. Vuyyuru, G.M., Talasila, D. and Marupudi, T.V., 2020, January. BRUKID Sort-A New Sorting Approach Using Multi Dimensional Array. In *2020 International Conference on Computer Communication and Informatics (ICCCI)* (pp. 1–6). IEEE.
9. Kopetz, H. and Steiner, W., 2022. *Real-time systems: design principles for distributed embedded applications*. Springer Nature.
10. Candra, A., Budiman, M.A. and Hartanto, K., 2020, July. Dijkstra's and a-star in finding the shortest path: a tutorial. In *2020 International Conference on Data Science, Artificial Intelligence, and Business Analytics (DATABIA)* (pp. 28–32). IEEE.
11. Palanisamy, V. and Vijayanathan, S., 2020, October. A novel agent based depth first search Algorithm. In *2020 IEEE 5th International Conference on Computing Communication and Automation (ICCCA)* (pp. 443–448). IEEE.
12. Porter, K., Nordvik, R., Toolan, F. and Axelsson, S., 2021. Timestamp prefix carving for filesystem metadata extraction. *Forensic Science International: Digital Investigation, 38*, p.301266.
13. Gommers, R., Virtanen, P., Burovski, E., Weckesser, W., Oliphant, T.E., Haberland, M., Cournapeau, D., Reddy, T., Peterson, P., Nelson, A. and Wilson, J., 2022. scipy/scipy: SciPy 1.9. 0. *Zenodo*.
14. Harris, C.R., Millman, K.J., Van Der Walt, S.J., Gommers, R., Virtanen, P., Cournapeau, D., Wieser, E., Taylor, J., Berg, S., Smith, N.J. and Kern, R., 2020. Array programming with NumPy. *Nature, 585*(7825), pp.357-362.
15. Vyas, S., Tyagi, R.K., Jain, C. and Sahu, S., 2021, July. Literature review: A comparative study of real time streaming technologies and apache kafka. In *2021 Fourth International Conference on Computational Intelligence and Communication Technologies (CCICT)* (pp. 146–153). IEEE.

16. Sholichah, R.J., Imrona, M. and Alamsyah, A., 2020, September. Performance analysis of Neo4j and MySQL databases using public policies decision making data. In *2020 7th International Conference on Information Technology, Computer, and Electrical Engineering (ICITACEE)* (pp. 152–157). IEEE.
17. Matallah, H., Belalem, G. and Bouamrane, K., 2020. Evaluation of nosql databases: Mongodb, cassandra, hbase, redis, couchbase, orientdb. *International Journal of Software Science and Computational Intelligence (IJSSCI)*, *12*(4), pp.71-91.
18. Ahmed, N., Barczak, A.L., Susnjak, T. and Rashid, M.A., 2020. A comprehensive performance analysis of Apache Hadoop and Apache Spark for large scale data sets using HiBench. *Journal of Big Data*, *7*(1), p.110.
19. Zaripov, R.N., Murakaev, I.M., Novikov, S.V. and Ryapukhin, A.V., 2020. Corporate structure for innovative enterprises. *Russian Engineering Research*, *40*, pp.137-139.
20. Axtmann, M., Witt, S., Ferizovic, D. and Sanders, P., 2022. Engineering in-place (shared-memory) sorting algorithms. *ACM Transactions on Parallel Computing*, *9*(1), pp.1-62.
21. Davitkova, A., Milchevski, E. and Michel, S., 2020, March. The ML-Index: A Multidimensional, Learned Index for Point, Range, and Nearest-Neighbor Queries. In *EDBT* (pp. 407–410).
22. Menz, M., Kunisch, S., Birkinshaw, J., Collis, D.J., Foss, N.J., Hoskisson, R.E. and Prescott, J.E., 2021. Corporate Strategy and the Theory of the Firm in the Digital Age. *Journal of Management Studies*, *58*(7), pp.1695-1720.
23. Demirgüç-Kunt A, Peria MS, Tressel T. The global financial crisis and the capital structure of firms: Was the impact more severe among SMEs and non-listed firms?. Journal of Corporate Finance. 2020 Feb 1;60:101514.
24. de Las Heras-Rosas, C., Herrera, J. and Rodríguez-Fernández, M., 2021. Organisational commitment in healthcare systems: A bibliometric analysis. *International Journal of Environmental Research and Public Health*, *18*(5), p.2271.
25. O'Connell, M., Barry, J., Hartigan, I., Cornally, N. and Saab, M.M., 2024. The impact of electronic and self-rostering systems on healthcare organisations and healthcare workers: A mixed-method systematic review. *Journal of Clinical Nursing*.
26. Sunyaev, A. and Sunyaev, A., 2020. Distributed ledger technology. *Internet computing: Principles of distributed systems and emerging internet-based technologies*, pp.265–299.

Workloads Scheduling in Scientific Workflow-as-a-Service Cloud Architectures

Victor Toporkov, Dmitry Yemelyanov, and Alexey Tselishchev

Abstract Workflow is currently the most common execution model for composite applications across multiple disciplines: seismology (CyberShake), bioinformatics (Epigenomics, SIPHT), astrophysics (Montage), gravitational wave physics (LIGO), hydro and aerodynamics, quantum chemistry, nanotechnology, hydrometeorology, modeling of social systems and transport infrastructure. The workflow is usually a collection of interrelated tasks within the directed acyclic graph (DAG) model parameterized by a priori, usually inaccurate, user estimates for tasks (relative computational or data transfer volumes, execution durations etc.). In this work, we propose an approach for scheduling science-intensive applications within the framework of the concept of Workflow-as-a-Service (WaaS). The proposed scheduling model is built based on the critical jobs' method, which allows us to obtain the deadlines for completing each of the workflow tasks under given efficiency criteria and inaccurate user estimates. This schedule must take into consideration the structure and dynamics of the WaaS resources' utilization and lifecycle of virtual machines (VMs). To solve this problem, we propose a novel procedure to group and assign workflow tasks to VMs instances provided by the Infrastructure as a Service (IaaS) provider.

Keywords Cloud computing · Scientific workflow · Scheduling · Inaccurate estimates · Critical job · Batch · Task · Hungarian algorithm

V. Toporkov (✉) · D. Yemelyanov · A. Tselishchev
National Research University, Moscow Power Engineering Institute (MPEI), Moscow, Russian Federation
e-mail: ToporkovVV@mpei.ru

D. Yemelyanov
e-mail: YemelyanovDM@mpei.ru

A. Tselishchev
e-mail: Alexey.Tselishchev@gmail.com

N. Polatidis et al. (eds.), *Intelligent Distributed Computing XVII*, Studies in Computational Intelligence 1203, https://doi.org/10.1007/978-3-031-87639-4_3

1 Introduction

Numerous prominent scientific projects, including CyberShake, Montage, SIPHT, Epigenomics, and LIGO, are executed as workflows [1–16]. Infrastructure as a Service (IaaS) enables a Workflow Management System (WMS) to access a virtually unlimited pool of resources on a "pay-per-use" basis. To date, a wide array of workflow management systems has been developed [17], including Kepler, ASKALON, CloudBus, Pegasus, Galaxy, HyperFlow, Taverna, among others. The Workflow as a Service (WaaS) paradigm facilitates the implementation of effective mechanisms for managing continuous streams of diverse job types within cloud computing environments. However, this paradigm also presents several fundamental challenges related to the organization and scheduling of heterogeneous job flows and composite applications.

The main aim of the proposed solution is to establish core algorithms for Workflow as a Service platform that enable the monitoring, processing of requests, scheduling, and management of heterogeneous cloud resources. This solution specifically emphasizes the dynamic creation and removal of virtual machines (VMs) and containers to optimize their allocation for various workflow tasks.

The proposed solution intends to take into account a number of important aspects: (1) the presence of multiple IaaS providers and different types of resources; (2) geographic distribution of data centers; (3) heterogeneity of workflows; (4) the need to implement the "pay-per-use" model for a specific system user; (5) finally, solving the problem of an efficient deploying of VMs and providing many containers ready for multi-threaded environment on physical servers.

The proposed solution aims to address the following main aspects: (1) the existence of multiple Infrastructure as a Service (IaaS) providers and various types of resources; (2) the geographic distribution of data centers; (3) the heterogeneity of workflows; (4) the necessity of implementing a "pay-per-use" model tailored to individual system users; and (5) the challenge of efficiently deploying virtual machines and data transfers in a multi-threaded cloud environment.

The rest of this paper is organized as follows. Section 2 reviews work that is related to our discussion. Section 3 presents the critical jobs' method (CJM) method for scheduling science-intensive applications and its implementation. Sections 4 and 5 introduce strategies and a general optimization scheme for VMs allocation. Sections 6 and 7 provide details on the software implementation and present the workflow scheduling results obtained for the algorithms under consideration. Section 8 summarizes the paper.

2 Related Works

The development of cloud technologies and WMS has given new impetus to research in the field of workflow scheduling in various applications. In particular, one of the areas of research is related to the active development of a paradigm in cloud computing WaaS [1–16].

As a rule, in known scheduling algorithms, the total cost of executing a workflow is used as one of the optimization criteria or restrictions [1–6, 13]. The work [1] proposes an approach to scheduling workflows in a container cloud environment for the WaaS model. The paper [2] introduces the use of a workflow broker based on a combination of on-demand and spot resource instances to minimize flow execution costs while meeting deadline constraints. The authors of [4] propose a time- and budget-aware dynamic workflow scheduling (DDBWS) algorithm designed specifically for WaaS environments. DDBWS schedules workflows by solving the problem of packing multiple resources. Unlike existing algorithms, it simultaneously considers the processor and memory demands of tasks. The proposed algorithm can significantly reduce the total number of rented VMs. In [6], the results of expanding the functionality of WMS CloudBus to process multiple workflows are presented and a prototype of a WaaS cloud platform for applications in the field of bioinformatics is proposed. A budget-constrained resource scheduling algorithm for multiple workflows (EBPSM) is implemented. In [13], a scheduling method is proposed that can reduce monetary costs and complete the work process within the minimum execution time. To analyze the performance of the proposed algorithm, an experiment is carried out in the WorkflowSim environment, and the results are compared with existing well-known algorithms—HEFT and DHEFT.

One of the main challenges of workflow scheduling is to make it energy efficient for cloud providers [3, 14, 15]. To date, formal formulations of workflow scheduling problems in scientific applications with several criteria are known [7, 8]. Artificial intelligence methods are increasingly being used in workflow scheduling tasks [5, 9, 10]. An important issue is representing the workflow model as a DAG. In a number of applications, loops are present naturally. In some known WMS (Pegasus. Apache Airflow, Taverna, Kepler) palliative techniques are used. This results in increased planning time.

In this work, in contrast to the studies discussed above, we propose innovative technologies and tools for scheduling and managing workflows of varying complexity and structure, considering many factors affecting the efficiency of using cloud platform resources, and the simultaneous passage of workflows on WaaS platforms.

3 Workflow Scheduling with Critical Jobs' Method

3.1 Critical Jobs' Method

The core of a Workflow as a Service (WaaS) system is undoubtedly the algorithm that manages the processing and scheduling of workflows. Although there are various approaches to tackle this issue, including both classical methods and those mentioned in the preceding section, we will commence our analysis with the Critical Jobs Method (CJM). A notable feature of CJM is its ability to produce a reference scheduling plan and set execution deadlines for each task within the workflow, relying on a priori estimates provided by users, which are frequently subject to imprecision.

More formally, the Critical Jobs Method algorithm addresses the following problem: the workflow of data-dependent tasks can be represented as a Directed Acyclic Graph (DAG), where the vertices correspond to tasks and data transfers. The processing of independent task flows is conducted in groups, with tasks arranged according to their priorities. A job is defined as a sequence of tasks, which corresponds to a path within the DAG. Let G be a parameterized workflow graph with various levels of parallelism of its partially ordered tasks. The partial order relation on the set $T = J \cup D$ of tasks and data transfers is defined with a DAG, where a subset J correspond to computational tasks, and a subset D—to the data transfers between the tasks. The set of oriented edges of the graph represents information and logical connections and dependencies. The graph is parameterized by a priori estimates of relative computational or data transfer volumes v_{ik}, execution durations t_{ik}^0 for tasks $j_i \in J, i = 1, ..., n$, on the corresponding resource type $k \in K$, where K is a number of resource types, and n is a number of tasks.

We define the distribution r of resources between tasks in J over a period of time $[0; t^*]$ as follows:

$$r = \{(a_i, b_i, t_i), i = 1, .., n, a_i = k \vee k^\circ, k = 1, .., K, k^\circ = 1, .., K, b_i \in [0; t^*]\}, \tag{1}$$

where a_i is a parameter determining the assignment of a task $j_i \in J$ to the corresponding resource; b_i and t_i are, respectively, the start time and execution duration of the task $j_i \in J$ on the resource, the type of which is determined by the assignment a_i.

In (1), $a_i = k$ if a task $j_i \in J$ is tied to a so-called base resource, the level (for example, the number of processors) of which is limited and depends on the capabilities of the task parallelization system, the cost of using the resource of the type k, and a number of other factors.

In the event of a conflict between parallel tasks from J, competing for the same resource of type k, taking into account the scalability of the computing environment, a resource of type $k^\circ \in \{1, .., K\}$ is introduced that is not inferior in its characteristics to the basic one, and at the same time $a_i = k^\circ$. This may be, for example, an additional processor node of the same type k or an unused base node of the type $k^\circ \neq k$.

Let us assume that restrictions (deadlines) are configured for completion times of individual tasks and jobs, i.e. sequences $j_{i1}, .., j_{iL}$ of informationally or logically related tasks $1 \leq i_1 \leq .. \leq i_L \leq n$ that make up the job:

$$t_g^* - t_g \geq 0, t_h^* - \sum_h t_h \geq 0, g, h \in \{1, .., n\}, \tag{2}$$

where t_g, t_h—completion times of tasks j_g, $j_h \in J$, and t_g^*, t_h^*—deadlines for task j_g and the job containing the task j_h.

One example of a scheduling criterion is the function of a workflow completion cost:

$$CF = \sum_{i=1}^{n} \lceil \frac{v_{ik}}{t_{ik}} \rceil, t_{ik} \geq t_{ik}^{0}, \tag{3}$$

where v_{ik}—is the relative computational volume of task j_i; t_{ik}—is the time allocated for executing a task j_i on a processor of type k; n—is the number of tasks; $\lceil \cdot \rceil$ denotes the nearest integer that is not less than the value $\frac{v_{ik}}{t_{ik}}$ of the cost of executing the ith task.

Resource allocation (1) is admissible if constraints (2) are met and the corresponding optimality criterion is defined, for example (3).

A **critical job** is a sequence of tasks (containing unassigned tasks) with the largest sum of specified prior execution estimates using the **best combination of resources**.

The iterative application of conflict resolution among parallel tasks that compete for the same resource, as outlined in (1), is fundamental to the Critical Jobs Method. Utilizing this method, it becomes feasible to develop a framework for the sequential creation of both reference (optimal) and suboptimal schedules based on a specified efficiency criterion. Let the criterion for the efficiency of resource use be given as a cost function (3). The CF takes the closest to t_{ik} duration estimate t_{ik}^{0} that determines the type k of resource being used (slot set or node).

Next, we highlight the primary attributes and constraints of the original CJM:

- The method is specifically designed for scheduling only a single workflow.
- The core principle of CJM is to implement a multi-phase procedure that identifies the subsequent critical job and resolves any potential conflicts (collisions) with tasks that have already been assigned to shared resources.
- It facilitates the producing of optimal or nearly optimal schedules while adhering to specified constraints related to total cost or task execution deadline in a single workflow.

3.2 Critical Jobs' Method Modification

In the considered dynamic workflow processing environment, the core CJM algorithm is designed to handle incoming workflow jobs on an individual basis as they arrive at the platform. A significant challenge that emerges is the system's dynamic nature and the need to adhere to specific quality of service requirements for each workflow. In the standard implementation of CJM, during the conflict resolution phase, a method is proposed to simplify the bipartite graph representing a "multi-layer" collision into a "two-layer" collision graph. This is achieved by systematically examining adjacent pairs of graph vertices (tasks) and selecting those with the appropriate weight. Consequently, each scheduling of a critical job has the potential to conflict with prior allocations, which can result in collisions among parallel tasks and jobs from multiple workflows being processed simultaneously.

In our dynamic model, we focus exclusively on the initial phase of the CJM for all incoming workflow jobs. This phase prioritizes the planning of time ranges for executing critical job tasks, ensuring compliance with user-defined budget and time constraints. Importantly, these calculations can be executed concurrently (in parallel) across various distinct workflows without facing conflicts for the resources. The resulting time ranges provide execution recommendations and deadlines for each task, which will subsequently be handled and processed by the cloud resource assignment module.

While developing the method, several assumptions have been made.

1. As the CJM planning stage considers only resource types rather than specific instances, there are no intersections or conflicts within a single job, a workflow jobs, or across multiple workflows. Consequently, the resource matching and assignment is deferred to the stage of assigning tasks to specific instances of VMs.
2. The modified CJM constructs an execution plan for a working based on a priori time estimates for task execution derived from the specified resource types, as well as the data transfer times between workflow tasks. Information regarding the volume of data transmitted and the duration of each data transaction is documented in the output of the CJM operation and subsequently forwarded to the assignment module.
3. During the scheduling stage of a critical job and the calculation of the additive separable criterion, if multiple identical values of the criterion are encountered, we opt for the first one. At this stage, it is crucial to select the optimal value without regard to the specific assignment of resource types.

As a result, the modified Critical Jobs Method is applied to each new workflow that arrives on the Workflow as a Service platform. The output includes the following: time ranges for the execution of workflow tasks, the volume and execution requirements of the tasks, as well as the data volumes and transfer dependencies between the tasks within the workflow.

4 The General Optimization Scheme of the Virtual Resources Allocation

4.1 Virtual Resources Allocation

A significant challenge in scheduling and executing a large number of computational tasks within workflows is the efficient allocation and management of virtual machines (VMs). This process entails determining the optimal timing for starting (creating) and stopping (releasing) individual VMs and containers, as well as establishing the assignment and execution order of tasks that are prepared to run. We propose several strategies for managing VMs to enhance the execution of workflow tasks.

One approach involves creating a specialized dedicated virtual machine for each individual task and releasing it upon the task's completion. This **greedy** yet flexible strategy allows for a precise alignment of VM types and lifetimes with the specific requirements of the tasks, especially when the time required to create and shut down a VM is considerably shorter than the average execution time of the tasks. However, the VM creation and preparation phase encompasses the time needed to configure the necessary software environment and transfer the required input data. Furthermore, the shutdown and cleanup time may increase if there is a need to save and transfer the results of the computations to the subsequent task or to a global data storage solution, such as Amazon S3.

Another approach involves maintaining a dynamic pool of continuously active virtual machines and distributing ready-to-run tasks among them, referred to as the **control strategy**. The dynamics of this pool entail adjusting the number of active machines in response to current computing demands. This method can facilitate more efficient scheduling and task assignment by aligning task requirements with the VMs that are already active, potentially allowing for the elimination of data transfer routines. For example, when two consecutive tasks that are dependent on one another are executed on the same VM, the need for copying and transferring data is removed. However, due to the diverse nature of workflow structures and their varying quantities, it is not always possible to ensure full and constant utilization of the entire pool of active VMs. Consequently, some VMs may remain idle at times, which can reduce the effectiveness and cost-efficiency of this strategy. Additionally, a challenge arises in developing an algorithm that can efficiently assign tasks to the available VMs.

The third approach is a mixed strategy that entails maintaining a base pool of permanently active VMs throughout the execution of workflows, while also permitting the creation of additional dedicated VMs to accommodate any unassigned tasks that are ready to run.

Considering that the implementation of the first greedy approach is relatively simple yet lacks effective mechanisms for optimizing the utilization of virtual resources, this paper will delve deeper into the control strategy.

4.2 High-Level Optimization Scheme

To implement the control strategy, we envision the following high-level stages.

1. The algorithm takes as input the execution time plans for individual tasks, the data transfer volumes between them, and the available types of virtual machines. The task execution plan specifies the expected time interval for actual execution, including the earliest start time and the latest completion time. It is assumed that these time ranges are provided by a modified implementation of the Critical Jobs Method (CJM), which preserves the continuity and sequence of execution within the original workflows.

2. In the initial stage, the input task flow is segmented into parallel execution batches. The primary criterion for this segmentation is that all tasks within a single batch must be independent, indicating that there are no data dependencies among them, which facilitates their concurrent execution according to the execution plan. Consequently, the complete set of tasks is organized into multiple consecutive groups or batches. Tasks that have data dependencies must be allocated to different sequential parallel execution batches, although they do not need to be adjacent. This batching process can be performed dynamically to accommodate the ongoing influx of tasks. For the overall optimization scheme, it is sufficient to work with two parallel execution batches.
3. During the second stage, the algorithm carries out sequential scheduling and assigns tasks for each batch to the virtual resources. To accomplish this, it addresses the minimum perfect matching problem, also known as the assignment problem, by utilizing the Hungarian (Kuhn-Munkres) algorithm [19].

5 Algorithms to Group Tasks into Parallel Batches

5.1 Generalization and Input Data

A critical initial step for efficient virtual machine allocation and processing is to determine the number of VMs required simultaneously at any given time. This number is affected by various factors, including the structure of the workflows, the relationships between tasks, their execution times, and the history of VM assignments. Consequently, it should be dynamically assessed during runtime. To facilitate this, we propose and evaluate two distinct algorithms for grouping workflow tasks into batches. Each batch should comprise tasks that can be executed in parallel without infringing upon data dependencies and local deadlines.

Firstly, we proposed and implemented the algorithm Follow The Leader (FTL) [19]. The main idea behind FTL is to preserve the following invariant: *given the batch start time and tasks' deadlines, no two tasks in the batch can be executed sequentially even on the fastest VM*. Consequently, the FTL establishes the minimum required level of parallelism for the incoming task flow at any given moment. The outcomes of this process can be utilized for predicting and dynamically managing multiple VMs.

The FTL algorithm takes as input a set of computational tasks t_i and a list of available VM types Vm_t^j. Each task t_i includes the following information: its computational volume Vp^i (which is necessary for predicting its execution time Te^i on each type of VM), the volume of input data V_{in}^i, the volume of output data V_{out}^i, and the precedence relationships with its parent and child tasks. Additionally, an expected execution interval is defined for each task during the preprocessing stage by the CJM module, specifying the earliest start time tr_i and the deadline for completion Td^i.

Figure 1 illustrates the Gantt chart for an instance of the FTL algorithm applied to a randomly generated task flow. The tasks are assigned for execution in batches, with the number of the parallel execution batch indicated to the right of each task body (for instance, Batch 2).

5.2 ASAP Algorithm

However, despite the implementation of the above-described invariant, FTL algorithm tends to save and minimize average demand for VMs and to postpone the execution of all tasks closer to their deadlines. These features may lead to an increase in the required VM performance and may result in parent and child tasks not being included in adjacent packages. In turn, this may prevent the reuse of execution data when performing parent and child tasks on the same VM instance. Thus, as an alternative to FTL we considered the more straightforward algorithm described in [6]. The basic idea of this algorithm is to execute each task as soon as possible when all parent tasks are finished. Each task is placed into the batch following the batch containing its last parent task. We call this algorithm ASAP. Thus, in contrast to FTL, ASAP strives to execute all tasks immediately when ready, without considering local deadlines. However, looking ahead, this policy of placing tasks into batches as early as possible turns out to be more suitable and flexible for the VM assignment stage.

Another important numerical metric of the batch grouping quality is the number N_{pc} of parent and child tasks in adjacent batches. Generally, such groupings may allow us to save time on data transfers due to reuse of VMs, their internal data storage and configurations. Table 1 shows N_{pc} provided by FTL and ASAP algorithms for different workflows. The results in Table 1 demonstrate the general advantage of the ASAP algorithm over FTL in terms of N_{pc} even for small workflows consisting of 50 tasks. On the other hand, the batch grouping result strongly depends on the workflow structure and the given task deadlines. The greatest advantage is demonstrated in

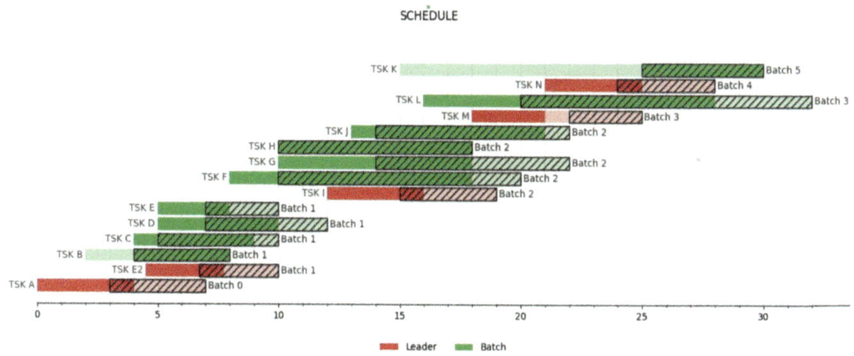

Fig. 1 Example of batch grouping with the FTL algorithm

Table 1 Number N_{pc} provided by FTL and ASAP algorithms for different workflows

Algorithm	LIGO50	GENOME50	CYBERSHAKE50	MONTAGE1000
FTL	31	15	29	834
ASAP	38	48	45	834

heterogeneous workflows with tasks of varying duration. However, for MONTAGE workflow consisting of 1000 tasks, ASAP and FTL algorithms provided identical batch groupings with $N_{pc} = 834$. As a result, ASAP algorithm generally demonstrates more efficient results because it groups tasks into batches more tightly and allows for resource reuse during the workflow execution.

5.3 Controlling the VM Allocation

Next, we focus on the effective allocation and assignment of resources for each parallel batch generated by the FTL algorithm. It is crucial to highlight that this assignment (VM matching) process naturally allows for the deactivation of unnecessary VMs. More formally, the problem for a given batch B can be expressed as follows:

1. Batch B contains N_B tasks ready for concurrent execution. For each task, the computational volume Vp^i, deadline Td^i and input data requirements (task and data dependencies) are predetermined.
2. The resource pool comprises N_{vm} active VMs that are ready to execute tasks from batch B. In general scenario $N_{\text{vm}} \neq N_B$. Furthermore, all VMs possess varying performance levels and may not be suitable for executing tasks from batch B, taking into account the specific tasks' requirements and deadlines.
3. It is essential to establish a mechanism for allocating new VMs and releasing excess VMs in order to efficiently execute a batch B of N_B tasks while minimizing the idle time of the active VMs.

We introduce the VM Allocation (VMA) algorithm to address the problem of efficient virtual machine allocation. This algorithm leverages the Hungarian (Kuhn-Munkres) algorithm (Kuhn, H. W., 1955) to identify a perfect matching within a bipartite graph $G = (T, R, E)$. In this context, T represents the set of tasks within batch B, while R denotes the set of available resources. The edges E connect the elements of T and R indicating that a task in T can be executed on a corresponding VM in R while satisfying all specified requirements. The weight assigned to each edge reflects a target optimization criterion for the assignment and scheduling process. Thus, the VMA algorithm enables the optimization of various factors, including the total cost of VM usage, the execution time of tasks, and the duration required for data transfers.

To achieve a perfect matching, it is necessary for the numbers of vertices in the subgraphs T and R to be equal. However, in general problem formulations, these numbers may differ ($N_{vm} \neq N_B$). To address this discrepancy, we propose the addition of pseudo-vertices to the sets T and R. For instance, we can introduce a vertex of type T_{off} to set T, which corresponds to the release of a VM when assigned to it. Alternatively, we can add a vertex of type R_{on} to set R, which triggers the creation of a new VM for the matching resource type. By appropriately adjusting the quantities and ratios of the additional T_{off} and R_{on} vertices, we can effectively configure and solve a subproblem related to resource allocation for each batch B (Fig. 2).

If the resulting match contains an edge directly connecting vertices of types T_{off} and R_{on}, the creation of the corresponding new VM resource is unnecessary.

The complexity of the proposed algorithm can be estimated as $O(N_B^3)$, where N_B represents the number of tasks in the parallel batch. This complexity is generally much lower than the total number of tasks present in the incoming workflows.

6 Software Implementation and Analysis

In the process of developing the software implementation for the proposed algorithms, several key assumptions were established.

Firstly, the current implementation operates under the assumption that the expected execution time of a task on a specific instance of a VM can be determined by calculating the ratio of the task's computational volume to a defined performance characteristic of the VM. This assumption is not mandatory; rather, it is introduced to improve the clarity of the model, the input data, and the associated calculations.

The second assumption posits the existence of a global data storage system with adequate capacity to concurrently accommodate all intermediate data necessary for transfers between tasks. Furthermore, it is assumed that the calculated data transfer rate remains constant during the parallel copying of output data from a set of completed tasks. This assumption is essential for simplifying calculations in scenarios where sequential tasks are executed with an acceptable delay, thereby making it more

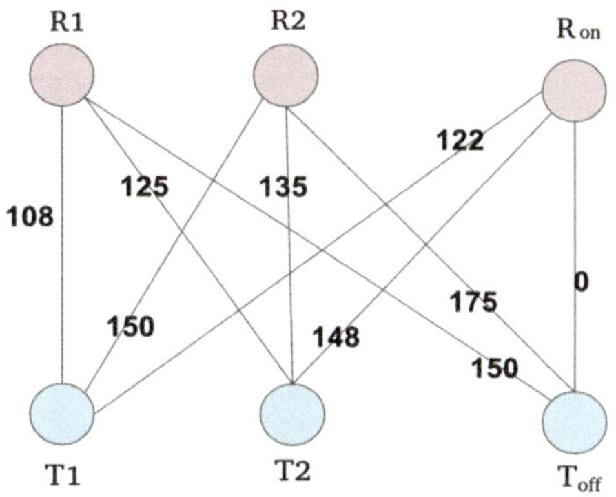

Fig. 2 Example of bipartite graph for the VM-task assignment

beneficial to transfer output data to centralized storage rather than retaining it on a VM until the subsequent task commences. This assumption is particularly valid in situations where the volume of transferred data does not exceed a specified critical threshold, and centralized, potentially cloud-based storage can effectively manage requests across multiple nodes.

7 Workflow Scheduling Optimization Results

7.1 Basic Optimization Results

To showcase the optimization capabilities of the proposed algorithm, we carried out a series of scheduling experiments on various real-world workflows, including CYBERSHAKE, GENOME, MONTAGE, LIGO, SIPHT, and their combinations provided as WaaS input.

Figure 3 illustrates an example of a LIGO workflow structure. Table 2 presents the key execution characteristics of the LIGO workflow across various optimization scenarios. The results demonstrate considerable optimization potential achieved by choosing the appropriate criterion in the considered algorithm.

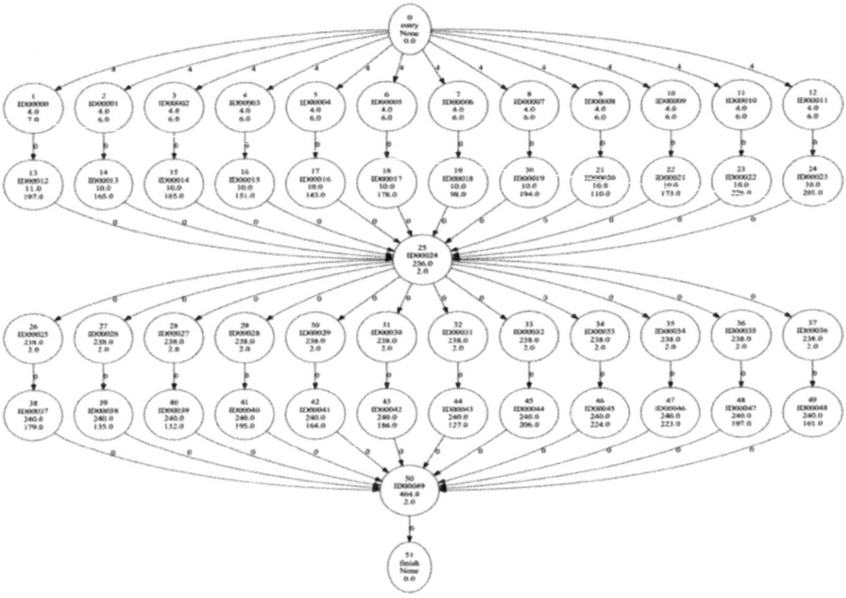

Fig. 3 Graphical representation of the LIGO workflow (Python 3, Graphviz)

Table 2 LIGO workflow optimization results

Optimization	Total VM cost	Total runtime, s	Total VM time, s
Runtime minimization	12929	**4180**	4310
Runtime maximization	12769	4757	4840
Cost minimization	**12740**	4260	4328
Cost maximization	**13057**	4576	4754
VM time minimization	12743	4200	**4269**
VM time maximization	12952	**4823**	**4980**

Thus, the cost minimization procedure yielded the lowest total execution cost for the workflow, while the runtime minimization criterion resulted in the shortest workflow runtime, and so forth.

7.2 Comparison of VM Allocation Strategies

In this experiment series we study VM allocation efficiency provided by the following strategies and algorithms.

1. **Greedy** strategy creates a new dedicated VM for each workflow task. VM type is selected to optimize the global scheduling criterion while meeting the execution time deadline. VM usage may require additional uptime to receive data transfers required for the task execution.
2. **Control** strategy is represented with the proposed VMA algorithm: VM assignment procedure optimizes global criteria given the deadline constraints for workflow tasks grouped in batches.

Testing was carried out based on a workload consisting of 100 independent workflows. The workflow instances were built based on real scientific applications (Montage, Cybershake, Genome, LIGO, SIPHT) and contain 50 vertices each. The execution deadline for each individual workflow was generated randomly between the fastest (using highest performance VMs) and the slowest (using the least performance VMs) possible execution times. The deadline limit determines the base execution plan and therefore affects the allocation variants and strategies. These workflows differ in their structure, required computational and data transfers volumes. For example, the average execution time of Montage workflow is 100 s, while Cybershake on average requires 30000 seconds. So, workflows with larger computational volumes may have a greater impact on the simulation results.

The following environment configuration parameters were studied:

- the arrival rate of workflows to the WaaS platform (quantity per minute);
- time intervals required to create and initialize and to release the VM.

Table 3 Algorithms' comparison depending on workflow arrival rate

Arrival rate	Algorithm	Execution time	VM cost	Created VMs
0.5	VMA	409280	13226	3912
1	VMA	409486	13277	4116
2	VMA	409602	13316	4334
6	VMA	409601	13279	4503
12	VMA	409586	13257	4574
60	VMA	409578	13168	4619
100	VMA	409596	13168	4633
**	Greedy	409650	13906	4955

It is worth to mention CJM scheduling step allows us to specify a specialized optimization criterion for each received workflow. This optimization will generally affect deadlines for the tasks of the workflow.

However, only one common optimization criterion can be used in VMA during the actual VM allocation. To support specific optimization criteria for input workflows it is possible to run several instances of VMA algorithms, each processing workflow matching one particular global criterion.

Table 3 contains total results of the workload execution depending on the workflow arrival rate (from 0.5 to 100 workflows per minute). Firstly, we note that Greedy execution results do not depend on the workflow arrival rate: all VMs are created and tailored for specific tasks, so the absolute start time of the task does not affect the choice of the VM type. On the other hand, VMA scheduling directly depends on the density of the incoming tasks. So, based on the workflow arrival rate and the composition of the parallel batches, VMA allocated 5–20% less VM instances by reusing them to execute several consecutive tasks. This strategy resulted in nearly 5% advantage over the Greedy strategy by the total VM cost criterion. Total tasks' execution time remained nearly constant as it depends on the pre-configured deadlines specific to each workflow and independent from the arrival rate.

Table 4 contains total results of the workload execution depending on the time required to create and initialize and to release (destroy) VM. This parameter affects the result of both VMA and Greedy algorithms, as total cost directly depends on total VM usage time, including periods of VM initialization and release.

Based on the results obtained, VMA allocated 16% less VM instances to execute the same amount of tasks, resulting in an up to 10% advantage over Greedy algorithms by the total cost criterion. Sa expected, the advantage increases with increasing VM creation and initialization time.

Overall, the following main conclusions can be drawn from the comparison results:

- both VMA and Greedy algorithms meet workflow deadlines in 100% of simulation experiments; total cost optimization implies approximate equality in terms of total time criterion;

Table 4 Algorithms' comparison depending on VM initialization and release time

VM Init/Release time	Algorithm	Execution time	VM cost	Created VMs
0/0	VMA	391361	10878	4175
0/0	Greedy	391559	10885	4955
10/1	VMA	391297	11009	4127
10/1	Greedy	391559	11053	4955
100/10	VMA	391449	12144	4125
100/10	Greedy	391559	12557	4955
300/30	VMA	391453	14576	4134
300/30	Greedy	391559	15899	4955
500/50	VMA	391413	17076	4118
500/50	Greedy	391559	19242	4955

- VMA exploits the possibility of reusing VMs, minimizing VMs initialization time and data transfer times between data-dependent tasks; in this way VMA at average allocates 10–15% less VM instances, resulting in 5% less total VM usage cost.

8 Conclusions

In this study, we proposed and implemented a multi-module procedure for scheduling and executing multiple independent workflows. To achieve this, several modifications of the Critical Job Method were developed, including the capability for time modeling to receive and schedule a set of workflows that are distributed over time.

The resource assignment stage enables the optimization of various global characteristics related to cloud resource utilization. The proposed solution manages the pool of active virtual machines by specifying the creation, preparation, utilization, data transfer, and shutdown intervals for each instance. The developed solution has been assessed using several examples of real-world workflows. It is important to reiterate that the results of this work were largely derived from an analysis of both classical and highly specialized optimization algorithms.

The primary limiting factor is the high (cubic) computational complexity of the solution in relation to the degree of parallelism in the incoming task flow [19]. As a result, there is a natural constraint on the size of workflows that can be scheduled within a reasonable timeframe. Future research will focus on addressing the complexity of scheduling algorithms in scalable Workflow as a Service platforms.

Acknowledgements The work was supported by the Russian Science Foundation (project No 22-21-00372; https://rscf.ru/en/project/22-21-00372).

References

1. Karmakar, K., Tarafdar, A., Das, R.K. et al. Cost-efficient Workflow as a Service using Containers. J Grid Computing 22, 40 (2024). https://doi.org/10.1007/s10723-024-09745-7
2. Taghavi, B., Zolfaghari, B. and Abrishami, S. A Cost-Efficient Workflow as a Service Broker Using On-demand and Spot Instances. J Grid Computing 21, 40 (2023). https://doi.org/10.1007/s10723-023-09676-9
3. Tarafdar, A., Karmakar, K., Khatua, S., Das, R.K.: Energy-efficient scheduling of deadline-sensitive and budget-constrained workflows in the cloud. In: International conference on distributed computing and internet technology. Springer, pp. 65-80 (2021)
4. Saeedizade, E., Ashtiani, M.: Ddbws: A dynamic deadline and budget-aware workflow scheduling algorithm in workflow-as-a-service environments. J. Supercomput. 77(12), 14525-14564 (2021)
5. Qin, Y., Wang, H., Yi, S., Li, X., Zhai, L.: An energy-aware scheduling algorithm for budget-constrained scientific workflows based on multi-objective reinforcement learning. J. Supercomput. 76(1), 455-480 (2020)
6. Muhammad H. Hilman, Maria A. Rodriguez, and Rajkumar Buyya. Workflow-as-a-Service Cloud Platform and Deployment of Bioinformatics Workflow Applications. Preprint. June 2020. 30 p. https://www.researchgate.net/scientific-contributions/Maria-A-Rodriguez-2114894132
7. Anurina Tarafdar, Kamalesh Karmakar, Rajib K Das, Sunirmal Khatua. Multi-criteria scheduling of scientific workflows in the Workflow as a Service platform Computers and Electrical Engineering Volume 105, 2023, 108458
8. Vincenzo De Maio a, Dragi Kimovski . Multi-objective scheduling of extreme data scientific workflows in Fog Future Generation Computer Systems. Volume 106, May 2020, Pages 171-184.
9. Huifang Li a, Yizhu Wang a, Jingwei Huang a, Yushun Fan. Mutation and dynamic objective-based farmland fertility algorithm for workflow scheduling in the cloud Journal of Parallel and Distributed Computing. Volume 164, June 2022, Pages 69-82. https://doi.org/10.1016/j.jpdc.2022.02.005
10. Yang Gu, Jian Cao, Shiyou Qian, Nengjun Zhu, Wei Guan. MANSOR: A module alignment method based on neighbor information for scientific workflow Concurrency and Computation: Practice and Experience Volume36, Issue10, May 2024 e7736. https://doi.org/10.1002/cpe.7736
11. Ahmad, Z., Nazir, B., Umer, A.: A fault-tolerant workflow management system with quality-of-service-aware scheduling for scientific workflows in cloud computing. Int. J. Commun. Syst. 34(1), 4649 (2021)
12. B Burkat, K., Pawlik, M., Balis, B., Malawski, M., Vahi, K., Rynge, M., da Silva, R.F., Deelman, E.: Serverless Containers-rising viable approach to Scientific Workflows. In: 17th International Conference on eScience (eScience). IEEE, pp. 40-49 (2021)
13. Karmakar, K., Das, R.K., Khatua, S.: Resource scheduling for tasks of a workflow in cloud environment. In: International conference on distributed computing and internet technology. Springer, pp. 214-226 (2020)
14. Ranjan, R., Thakur, I.S., Aujla, G.S., Kumar, N., Zomaya, A.Y.: Energy-efficient workflow scheduling using container-based virtualization in software-defined data centers. IEEE Trans. Ind. Inform. 16(12), 7646-7657 (2020)
15. Silva, R.F., Pottier, L., Coleman, T., Deelman, E., Casanova, H.: Workflowhub: community framework for enabling scientific workflow research and development. In: 2020 IEEE/ACM workflows in support of large-scale science (WORKS). IEEE, pp. 49-56 (2020)
16. Medara, R., Singh, R.S. A Review on Energy-Aware Scheduling Techniques for Workflows in IaaS Clouds. Wireless Pers Commun 125, 1545-1584 (2022). https://doi.org/10.1007/s11277-022-09621-1

17. Peter Amstutz, Maxim Mikheev, Michael R. Crusoe, Nebojša Tijanić, Samuel Lampa, et al. (2024): Existing Workflow systems. Common Workflow Language wiki, GitHub.https://s.apache.org/existing-workflow-systems updated 2024-08-18, accessed 2024-08-18.
18. Victor Toporkov and Dmitry Yemelyanov (2021). Micro-scheduling for Dependable Resources Allocation In: Performance Evaluation Models for Distributed Service Networks. Studies in Systems, Decision and Control. Vol. 343. Editors: Bocewicz, Grzegorz, Pempera, Jarosław, Toporkov, Victor. Springer International Publishing., pp. 81-105.
19. Toporkov, V., Yemelyanov, D., Bulkhak, A., Pirogova, M. (2024). Job Batch Scheduling in Workflow-as-a-Service Platforms. In: Sokolinsky, L., Zymbler, M., Voevodin, V., Dongarra, J. (eds) Parallel Computational Technologies. PCT 2024. Communications in Computer and Information Science, Springer, Cham. vol 2241. Pp. 65-79.

Hybrid Approach Using LSTM and Gradient Boosting for Accurate and Robust Store Sales Forecasting

Shalim Sadman and Doina Logofătu

Abstract Sales forecasting has become strategically very specific to the companies operational efficiency and a customer's good experience in today's retail analytics evolvement. In this work, I am going to dig into the effectiveness of a hybrid model that made up by the long short term memories(LSTM) holding pattern in time and the idea of XGBoost which is feature handling. The proposal entails an initiative to improve the precision and effectiveness of store sales forecasting. The aim of this study is to compare the predictability of the XGBoost, LSTM, and a mixture model using a multitude of dataset containing sales records data with a variety of characteristics. A modelling is designed where multiple performances metrics such as Root Mean Squared Logarithmic Error (RMSLE), Mean Absolute Error (MAE), Root Mean Squared Error (RMSE), and the coefficient of determination (R^2) are used for performance comparison. The hybrid model however is a mixture between LSTM, which can handle sequential data very well and XGBoost that is known as a very efficient and quick feature analysis tool and is able to outdo the current sales forecasting as it stands. The research also shows that ensemble is an approach that tackles the weakness of single model but it forms the basis for the development of retail analytics using hybrid machine learning techniques.

Keywords Sales forecasting · XGBoost · LSTM · Hybrid model · Time series analysis · Feature importance

S. Sadman (✉) · D. Logofătu
Frankfurt University of Applied Sciences, Frankfurt am Main, Germany
e-mail: shalim.sadman@stud.fra-uas.de

D. Logofătu
e-mail: logofatu@fb2.fra-uas.de

N. Polatidis et al. (eds.), *Intelligent Distributed Computing XVII*, Studies in Computational Intelligence 1203, https://doi.org/10.1007/978-3-031-87639-4_4

1 Introduction

The sell off between retail supply chain efficiency and customer satisfaction in the modern world is barely to an exorbitant extent of one very important role of sales forecasting as [1, 15]. Accurate forecasting sales as a forerunner is paramount for companies in an era of technology, unruly consumers, and economic instabilities that they deal with tactically. The growing role of analytical skill is not only gain highly respectful but also is the source of the new era which is wholly based on digitalization in the realm of retail. This research employs the evidence-based machine learning just recently developed to upgrade the technology applied for prediction [1].

The swift development of the retail industry, driven by the utilization of big data and analytical developments, emphasizes the crucial requirement for accurate sales forecasting in order to adjust to consumer trends and economic fluctuations. Nevertheless, the intricate nature of contemporary retail data poses difficulties for conventional forecasting techniques, as they find it challenging to account for the non-linear impacts of consumer behavior[11].

Conventional time-series models like ARIMA, although helpful, typically struggle to handle the complex dynamics of the modern retail industry, characterized by factors like promotions and seasonality[5, 9]. Machine learning technologies, such as neural networks and deep learning algorithms, are powerful tools for detecting patterns in intricate datasets. LSTM networks, in particular, are highly regarded for their ability to accurately forecast time-series data [8, 17].

The combination of LSTM's deep learning capabilities with the predictive power of Gradient Boosting models in a hybrid approach holds the potential to overcome the limitations of traditional and standalone machine learning approaches. The purpose of this hybrid model is to improve the accuracy and adaptability of forecasting, specifically to tackle the constantly changing issues faced by the retail sector [4].

The objective of this project is to develop a sophisticated hybrid machine learning framework that will enhance the accuracy of sales predictions in retail stores. The undertaking is based on the need for retail businesses to effectively utilize analytical tools in a market environment that is focused on data [13]. The main goals are to create and apply a combined model that combines LSTM and Gradient Boosting, thoroughly evaluate its performance compared to individual models, and analyze its practicality in the context of supermarket sales forecasting, which has specific and complex requirements.

This paper focuses on improving the accuracy of supermarket sales forecasting by using a combination of LSTM and Gradient Boosting models. This study aims to explore the temporal patterns in sales data, enhance the incorporation of external variables into the model, and assess the effectiveness and applicability of the hybrid model in different retail situations [18].

This paper seeks to present the main advancements in theoretical knowledge, methodological creativity, and practical implementation in retail forecasting through machine learning. The aim is to improve the understanding of engineering hybrid models for complex forecasting tasks, provide a systematic methodology for implementing machine learning strategies, set performance standards, propose a framework for practical application and highlight the strategic implications for retail management.

2 Previous Work

The progress of machine learning (ML) technology has had a substantial influence on the retail industry, namely in the area of sales forecasting. This is achieved by utilizing extensive datasets on sales, customer behavior, and operational variables. This review compiles significant findings from multiple studies to emphasize the impact of machine learning (ML) in comprehending consumer behavior, expanding the accuracy of sales prediction, and improving operational efficiency in the retail industry.

Alawadh and Barnawi (2024) highlighted the significance of using many data sources, such as transactional data, customer demographics, and environmental factors, in order to properly predict consumer purchasing patterns using machine learning frameworks [3]. In a study conducted by Merdas and Mousa (2023), it was shown that the integration of sophisticated algorithms like as MLP, RANSAC, and Bagging can effectively deal with the inherent unpredictability found in supermarket sales data[12].

Within the realm of fresh food retail, the utilization of LSTM, Transformers, and XGBoost models has demonstrated significant achievements in predicting demand, thereby enhancing inventory management and minimizing wastage. In the study conducted by Guru (2023), the effectiveness of different machine learning models, including K-NN, Linear, Polynomial, and Ridge regression, was examined in terms of their ability to forecast product sales. The results emphasized that K-NN exhibited greater predictive performance compared to the other models [7].

The transition from conventional statistical techniques to advanced machine learning (ML) and deep learning models, such as Decision Trees, Random Forests, and neural networks, represents a notable change in the approach to analyzing intricate patterns in retail data [14]. The integration of LSTM networks with CNNs in deep learning has revolutionized time series forecasting by efficiently handling long-term dependencies and improving pattern recognition capabilities [10].

Hybrid models, which combine techniques such as LightGBM and LSTM, have become effective tools for improving the accuracy of sales forecasting. This highlights the potential of merging diverse machine learning techniques to take use of their distinct strengths [16]. Research conducted by Abhinaya et al. (2023) and Aggarwal (2023) demonstrates the wide range of uses for machine learning (ML), such as predicting supermarket sales using XGBoost and improving the online grocery shopping experience using AI and ML-based recommendations [2, 6].

3 Methodology

This paper's Methodology section provides a step-by-step strategy that was implemented to examine the efficiency of a hybrid machine learning model in predicting shop sales in line with the information received from the literature review. This chapter gives a comprehensive definition of the steps of the research process from data collection to model evaluation. It emphasizes the employment of stringent protocols to ensure the validity and precision of the study results. This research project seeks to associate theoretical developments in machine learning with practical usage in sales forecasting in the retail industry by systematically structuring the research framework. Hence, it will benefit both the theoretical and practical sides.

3.1 Research Design

I propose this work as a case study of a quantitative research that involves the use of machine learning methods in the analysis of sales data history. The research primarily focuses on the development and comparison of three models: using an LSTM model, independent by itself and powerful in processing time series data; an XGBoost model, member of the efficient family, having the ability to make classification and handling of the feature as a specialty; and finally a hybrid model which takes advantage of both approaches. The purpose is to come up with an answer; whether the hybrid model gives a significant increment to the accuracy of predictions compared to each of the individual models. For this exercise, theory is not the end, as it intends to find a solution for the business issues in the field of retail sales forecasting that can be used by predictive analytics practitioners.

3.2 Framework Overview and Hybrid Model Rationale

The article introduces a novel hybrid forecasting approach which leverages the strengths of Long Short-Term Memory (LSTM) networks and eXtreme Gradient Boosting (XGBoost) in providing accurate estimates of retail sales. The hybrid model integrates the power of LSTM to detect temporal patterns with the abilities of XGBoost to manage feature interaction and importance. The reason for employing this hybrid technique is rooted in its potential to enhance the precision of prediction by capturing both time-dependent patterns and complex correlations between several features in the sales data.

This approach relies on the concept that an ensemble model can produce a better predictive power if it combines the strengths of its member models. The recurrent architecture of LSTM network enables them to find patters that occur at different time scales and hence, they are effective for modeling time series data. Gradient Boosting

Table 1 Overview of methodological stages

Stage	Description	Purpose
Data Preprocessing	Normalization, handling missing values, and temporal feature engineering	To optimize model performance by preparing and cleaning the dataset
LSTM model training	Training the LSTM model on preprocessed time-series data	To analyze temporal connections and trends in sales data
Gradient boosting training	Training the Gradient Boosting model on the same dataset	To understand intricate feature relationships and improve predictions by minimizing errors.
Model predictions ensemble	Combining the predictions from LSTM and Gradient Boosting models through averaging or weighting	To capitalize on the characteristics of both models to enhance overall forecast accuracy
Performance evaluation	Applying metrics such as RMSLE, MAE, RMSE, and R2 to assess the hybrid model's forecasts	To statistically assess the predicting accuracy of the hybrid model and compare it to the performance of the separate models

improves predictions by sequentially improving the model and reducing errors in the process, by learning from the errors of the previous trees. The hybridization is aimed at addressing the limitations of individual models in handling the complex issues of sales forecasting. The purpose is to develop an extremely resilient and versatile forecasting tool that would be able to forecast outcomes adequately in all market situations and periods.

Table-1 acts as a brief guide into the different yet related stages of development of the hybrid model. Every step is very important, being based on previous actions and leading to formation of a powerful and flexible forecasting model that can successfully predict tendencies of the retail sales. The unification of LSTM and Gradient Boosting under a single forecasting platform is a breakthrough in predictive analytics. The paper offers an innovative approach that overcomes current limitations in sales forecasting and opens the door for further research in hybrid predictive modeling.

3.3 Data Sourcing

The study collected a complete dataset from a resource referred to as 'Store Sales - Time Series'. This dataset is an aggregation of multiple CSV files, each serving a unique purpose in the analysis:This dataset is an aggregation of multiple CSV files, each serving a unique purpose in the analysis:

- `train.csv`: The collection encompasses vital historical sales data essential for training the predictive algorithms.
- `test.csv`: Designated for evaluating the prediction capacity of the algorithms.
- `oil.csv`: Describing the correlation between oil prices and their possible influence on sales dynamics.
- `sample_submission.csv`: A standardized format for submitting forecasts about future events.
- `holidays_events.csv`: Aggregating holiday activities that could influence sales patterns.
- `stores.csv`: Providing a detailed description of the distinct characteristics of each store, such as its classification and geographical position.
- `transactions.csv`: Monitoring the number of transactions per store as a possible statistic for assessing sales effectiveness.

Each file was meticulously chosen to offer a distinct viewpoint in understanding the intricate nature of store sales.

3.4 Data Collection and Preparation

Dataset Description The dataset used entails daily sales data from different retail companies, which cover attributes such as sales figure, promotions, store type, and external economic indicators. The dataset spans different years, providing a broad temporal range for predictive research.

Data Cleaning and Preprocessing The first words of data processing are the elimination of the missing values, detection of outliers, and standardization of numeric variables in order to make them compatible with the model. Categorical variables include store type and promotions are one-hot encoded so that they are easily available for use in machine learning models.

3.5 Feature Engineering

Feature engineering is a key activity in implementing machine learning models by marrying two aspects: artistic as well as scientific. This study is a detailed procedure of selecting and modifying attributes that will capture the basic patterns and dynamics found in the sale's data. The models feature temporal elements which include lagged sales, promotions, and public holidays, which in turn improve the forecasting due to the holistic context they provide. This method makes use of both domain knowledge and literature research concepts to guarantee that the chosen features are theoretically well founded and empirically verified for their predictive power.

3.6 Model Implementation and Evaluation

LSTM Model The LSTM model is developed to capture the long-term dependencies of the sales data, with hyperparameters like the number of layers and neurons that have been optimized on the validation set. The model's architecture is specifically created for dealing with data that is sequenced, making it very appropriate for the tasks of predicting future events that contain time-related patterns. The structure of the LSTM model is carefully designed to handle the sequential nature of sales data, especially paying attention to seasonal patterns, and promotional influences that are important for retail forecasting. To solve overfitting, the model uses dropout layers, which makes sure that the model can generalize well to unseen data. The chosen activation function (commonly ReLU and tanh for hidden layers, and sigmoid for the output layer) is crucial in describing non-linear relationships in a model.
XGBoost Model The reason for using XGBoost is the efficiency of this approach at feature selection and high-dimensional data management. The model is tuned for parameters like learning rate, tree depth, number of estimators which help improve its predictive accuracy and minimize overfitting. One of the important features of XGBoost is its ability to handle the sparse data and complex feature interactions using gradient boosting. The fidelity of the process depends on fine tuning of hyperparameters which include the maximum depth of trees to control model complexity, and the learning rate that can balance speed and learning accuracy. Overfitting is efficiently blamed by the XGBoost model through the use of regularization parameters, which are then adjusted based on the cross-validation results.
Hybrid Model Design The hybrid model combines the LSTM and XGBoost models through a two-stage approach: LSTM model is used to detect the temporal patterns in the data. The LSTM model outputs, together with other features, are used as inputs to the XGBoost model for its feature interaction capabilities. This architecture tries to combine the benefits of both approaches in order to increase forecast accuracy. The hybrid model architecture smartly includes the LSTM model outputs with classic features, which are produced during the feature engineering step, and utilizes these as input for the XGBoost model. This method used by the model is tiered where time patterns are understood first, and large feature interactions are then analysed. This allows the model to adequately capture also the time dependent influences on sales. The interpretability of the model is meticulously focused on making sure that the results provide the actionable insights, which are very important for wise decisions in retail.

3.7 Model Training and Validation

The models are trained with a part of the dataset set aside for training. The models' performance is tested on an independent validation set to facilitate the tuning of hyperparameters and prevent the models from overfitting on the training data. Cross-validation procedures are used to assure the robustness of the model evaluation. The training approach utilizes a time-series cross-validation technique, that considers the sequential pattern of the data and assures of authentic and real-world manner of predicting future sales validation. The models are tested with different validation sets to fine-tune hyperparameters, with an emphasis on minimizing the discrepancy between training and validation performance to ensure that generalization will be reliable.

3.8 Performance Evaluation

The performance of the models in predicting the values is also measured in terms of metrics such as RMSE, MAE, and R^2, among others the three are widely applied in the realm of machine learning comprehensive evaluation instruments. A thorough analysis of a single model and mixing models' performance will facilitate selection of the best forecast performance.

4 Experimental Results and Tests

4.1 Model Performance Overview

The assessment of model performance focused on several crucial metrics: Root Mean Squared Error (RMSE), Mean Absolute Error (MAE), and R-squared (R^2). The measures provided a comprehensive view of the correctness and predictive capability of each model.

4.2 LSTM Model Evaluation

The LSTM model exhibited substantial proficiency in capturing the temporal relationships included in the sales data. The LSTM model, which has been designed with a specialized architecture for sequential pattern recognition, has obtained the following performance metrics (Table 2):

The obtained results demonstrate the ability of LSTM to work with time-series data, in particular, capturing long-term patterns and seasonal effects. However, performance of the model showed limitations in dealing with complex interactions of features and relevant external variables impacting the sales.

Table 2 Performance of the LSTM model

Metric	Value	Interpretation
MAE	292.597	Average error in the same unit as the sales figures
RMSE	701.854	Larger errors are more pronounced due to squaring
MAPE	0.1312	Errors are 13.12% of the actual sales on average
R^2	0.731779	Model explains 73.1779% of the variance in sales data

4.3 *Visualization of Model Predictions*

Several graphs were produced to represent the forecasted accuracy of the LSTM model. The graphs compared the forecasted sales of the model with the actuals sales data. The visualizations depicted performance of the model at different time points, captured times of good estimated accuracy and hinted about areas where the model could be improved through modifications.

Figure 1 gives a comprehensive picture of the predictive accuracy of the LSTM model throughout the lifetime of the dataset. The graph shows that the model is able to track the general sales direction, in particular by detecting inconstancies, which could mean possible improvements of the model.

Figure 2 shows a higher level of complexity. Figure 2 performs a detailed analysis of sales trends for 2017 and shows how the LSTM method follows these variations with an exceptional accuracy in a monthly manner. The strong correlation between the forecasted and actual sales data shows the model's capability to adjust the seasonal patterns and inherent changes in the sales data.

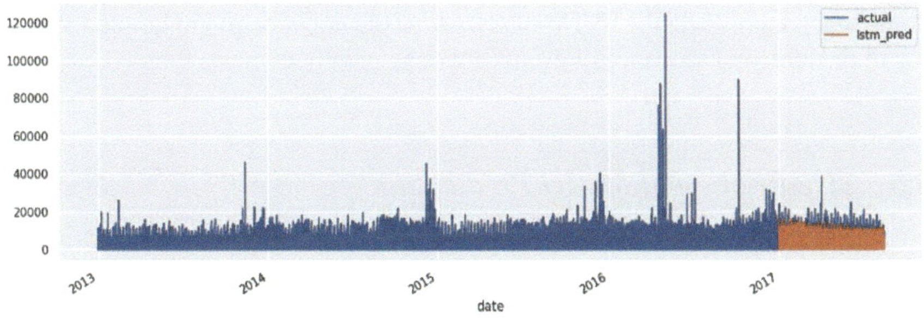

Fig. 1 Overall actual versus predicted sales by LSTM model

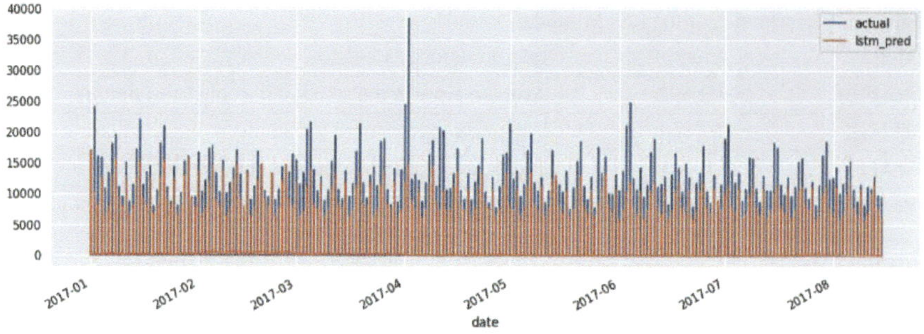

Fig. 2 Monthly actual versus predicted sales for 2017 by LSTM model

Table 3 XGBoost model performance metrics

Metric	Value	Description
RMSLE	0.00979517	Lower values are better
MAE	127.959	Error in the same unit as the target
RMSE	445.6475	Higher sensitivity to larger errors
R^2	0.891861	The model explains 89.186% variance

4.4 *XGBoost Model Evaluation*

XGBoost showed a great performance in feature selection and was able to handle high-dimensional data effectively that reflects its efficiency in the analysis of different retail data sets (Table 3). undefined

XGBoost model was better in the cases with complicated feature dependencies and non-linear patterns over the LSTM model. But it did not have a complete understanding of the temporal sequence of events affecting greatly the sales forecast accuracy.

4.5 *Visualization of Model Predictions*

The predicted results for the validation set were shown in a graphical form to allow a comparison between the projected and real sales data. The broad presentation included the whole data set, providing a comprehensive picture of the model's performance during the period. A complex visualization was created to look at the monthly sales data only for 2017. This graph goal was to demonstrate the model's capability to correctly identify sales trends over time, which include, seasonality, and unusual events.

Figure 3 provides an exhaustive representation of the discrepancy between forecast sales and actual sales throughout the period. The model nicely demonstrates that it

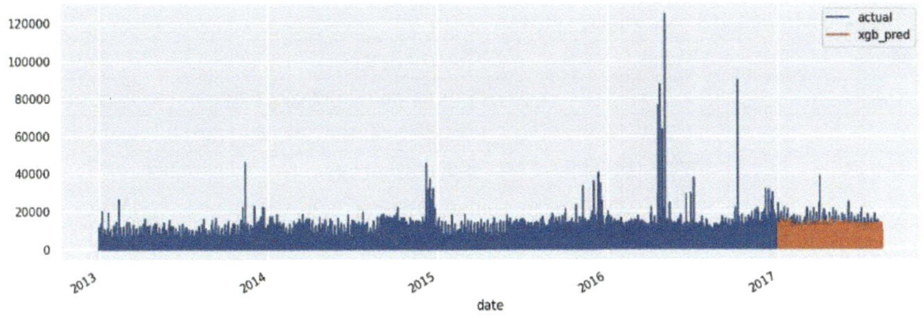

Fig. 3 Overall actual versus predicted sales comparison

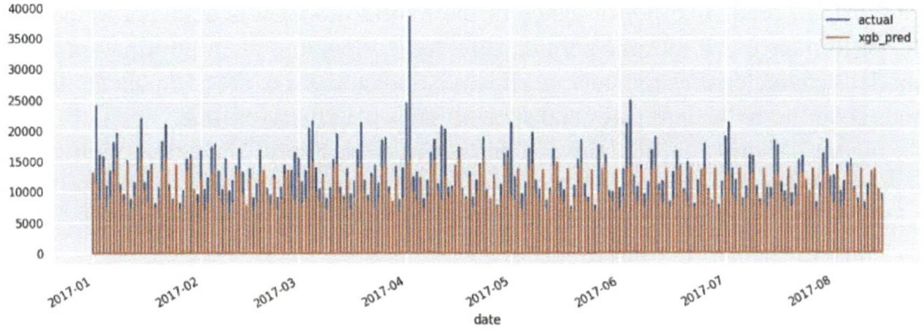

Fig. 4 Detailed monthly actual versus predicted sales for 2017

is able to capture sales trends in a consistent manner and also do a good job at identifying periods when there is a big departure.

Figure 4 in the monthly comparison for 2017 illustrates a strong correlation between the forecasted and actual sales, which indicates that the model has a good knowledge of time pattern sales and is tolerant of deviation in trends.

4.6 Hybrid Model Performance

The architecture of the hybrid model, which derives from LSTM and XGBoost models, is aimed at benefiting from the strongest sides of the both models (Table 4).

This model demonstrated better capability to address both temporal changes and the complex relationships between different features, which in turn, improves the prediction accuracy greatly compared to the individual models. The hybrid model efficiency indicates the synergistic power of the combination of LSTM and XGBoost for sales forecasting.

Table 4 Hybrid model performance metrics

Metric	Value	Description
RMSLE	1.780535	Lower values indicate better model accuracy
MAE	193.914729	Average absolute deviation from actual sales
RMSE	5376.05024	Sensitive to larger errors in predictions
R^2	0.842628	Indicates 84.2628% of variance is explained

4.7 Visualization of Combined Model Predictions

The overall predictive performance of the hybrid model was depicted to evaluate its performance in a set time period. This study offers important information on the model's characteristics to integrate and utilise advantages of the component models' thus covering both aggregates and specific sales patterns over time.

Figure 5 shows the predictive robustness of the hybrid model by plotting the actual sales in relation to the projected sales throughout the whole dataset. This visualization shows improved accuracy of the forecast and capability of the model to adjust to changes in sales trends.

The Fig. 6 provides a complete analysis of the monthly forecasts produced by the hybrid model for 2017. The model's ability to mirror sales volatilities and its adaptability to seasonal shifts is illustrated.

4.8 Comparative Analysis

The cross-sectional comparison of the three models showed that the model using hybrid data received lower score of RMSE and MAE, and had higher R^2 value. The comparison results of the alternative modeling approach enabled me to understand that it is possible to improve the reliability of forecasting with a combination of

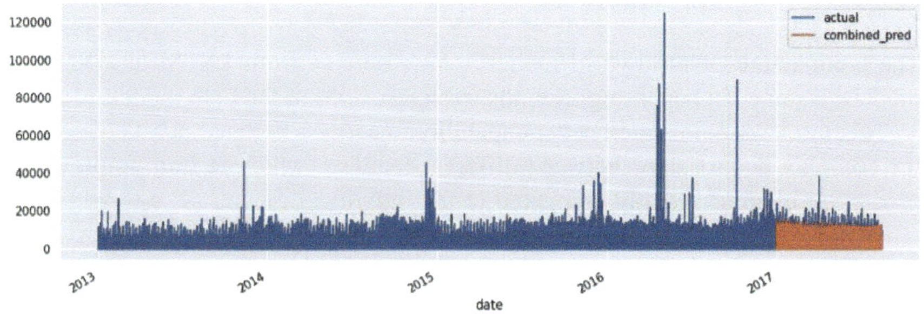

Fig. 5 Combined predictive performance of XGBoost and LSTM models

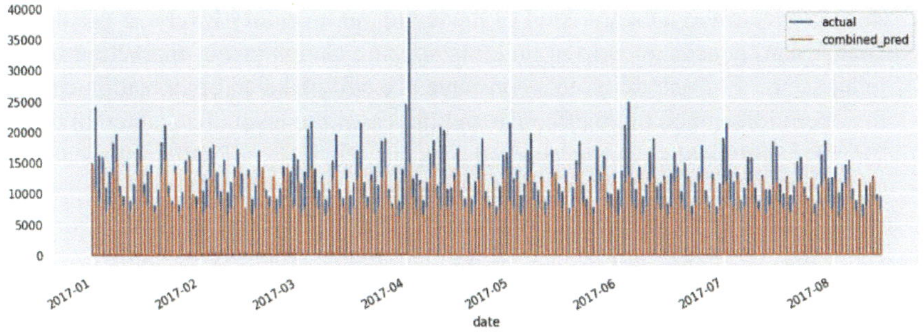

Fig. 6 Detailed monthly predictions for 2017 by the combined model

methods. Apart from this, the analysis was also an exercise in finding out the practical delivery strengths and weaknesses of the concept models.

- **RMSE Comparison:** Hybrid Model < XGBoost < LSTM
- **MAE Comparison:** Hybrid Model < XGBoost < LSTM
- **R^2 Comparison:** Hybrid Model > XGBoost > LSTM

4.9 Practical Implications of Findings

The facts emerging from this research have profound effects on the accuracy models forecasting the retail sales level. The performance of the hybrid model in all experiments shows a pragmatic solution to increase the forecasting accuracy. Having the accurate inventory levels, improved demand planning and strategy are the other benefits of using this hybrid approach. Moreover, the model's capability of being adaptable to any retail location and of being expand is an advantage in spreading this in the industry.

5 Discussion

This work demonstrates the strong performance achieved by a hybrid model by combining Long Short-Term Memory (LSTM) and eXtreme Gradient Boosting (XGBoost) algorithms that allows to depict retail sales with accuracy. Through such team procedure, I anticipate to exceed both the models' capabilities combined, giving us a superb tool to grasp market trends and sales dynamics. The hybrid architecture is a winner in temporal sequences recognition due to the use of LSTM and identification of complex feature relations of XGBoost what in its turn contributes to highly predictive accuracy.

This practice gives a clear vision of managing an inventory level and predicting a demand plan, hence can help in building specific campaigns of marketing too in the retail sector. For one thing, this consequently will make an organization carried out the operations much more effective and increases the level of satisfaction of the consumers. It solves several retailing industry challenges by allowing sales patterns as well as stock run-outs to be scrutinized through, which helps to develop stock harmony. Even if so, the effectiveness of this model is highly dependent on the quality of input data. Here the significance of the data management is strengthened. The largest barrier to its implementation among smaller businesses consists of computing requirements and it obliges a developer to perform an adjustment of a problem-solution for different scales.

The conclusions of our study are in agreement with the previous researches establishing the benefits of the combination machine learning models, and confirm the resourcefulness of applying this technology in retail sector. Seeing that the model mathematically verifies the application of many machine learning algorithms as well as the practicalness of the framework for detecting retail industry trends makes the study worthwhile. For retailers wanting to embrace this approach, their effort should start with building a well-functioning data system that will not get out of sync due to the shifts happening on the market.

The other factors include online sentiment and macroeconomic statistics, which could help the model become even more accurate. Further research may also study the flexibility of the model in other retail contexts, product types, and geographic areas. Continuous development of machine learning can make it more efficient and accurate in predicting, thus providing an opportunity for wider application of it in retail and other areas.

6 Conclusion and Future Work

The aim of this study is evaluating the efficacy of an ensemble machine learning model that incorporates LSTM and XGBoost for a forecast of a retail sales company. It illustrates that the proposed model integrates models and thus has higher accuracy rate than individual models. The contribution of this study is to give practical outcomes in theory, by showing that the combined structure of LSTM and XGBoost is able to provide an integrated model for shopping forecasting that goes beyond previous approaches reported in the literature. Accordingly, it provides the merchants with the platform that allows them to easily integrate the state-of-the-art machine learning methods with the system that actually enhances the precision of their predictions. From this research, though, I notice some limitations such as those of expertise data and enormous computer facilities. It points out the possible extension of such modeling like when to add new algorithms and also mentioning the challenges related to the implementation. This study can be seen as the point were the advances in

AI-application in the field of retail forecasting are merged with the practical requirements of the industry. This is a clear indication of the fact that forecasting models are gaining importance in the retail sector where things are constantly changing.

References

1. Abdul, A., Amin, A., Al Hasan, M.: Challenges and opportunities in big data analysis for digital marketing: a systematic review. Journal of Big Data **7**(1), 1–31 (2020). https://doi.org/10.1186/s40537-020-00329-2, https://journalofbigdata.springeropen.com/articles/10.1186/s40537-020-00329-2
2. Aggarwal, D.: Exploring the role of ai for online grocery shopping through enhancing personalized recommendations and customer segmentation. Technoarete Transactions on Advances in Computer Applications **2**(3) (Sep 2023), https://bit.ly/example-url, accessed: 2024-03-15
3. Alawadh, M., Barnawi, A.: A consumer behavior analysis framework toward improving market performance indicators: Saudi's retail sector as a case study. J. Theor. Appl. Electron. Commer. Res. **19**(1), 152–171 (2024), https://www.mdpi.com/0718-1876/19/1/9
4. Chen, H., Lundberg, S., Lee, S.I.: Hybrid gradient boosting trees and neural networks for forecasting operating room data. University of Washington (2020), https://ar5iv.labs.arxiv.org/html/1801.07384
5. Ding, C., Duan, J., Zhang, Y., Wu, X., Yu, G.: Using an arima-garch modeling approach to improve subway short-term ridership forecasting accounting for dynamic volatility. IEEE Transactions on Intelligent Transportation Systems **19**, 1054–1064 (2018), https://api.semanticscholar.org/CorpusID:4560010
6. Durga Abhinaya, C.S., Lahari, B., Priya, C.D., Anjali, D.: Supermarket sales prediction using machine learning. EPRA International Journal of Research and Development (IJRD) (Nov 2023https://doi.org/10.36713/epra14814, accessed: 2024-03-12
7. Guru, P., Sathyapriya, J., Rajandran, K.V.R., Bhuvaneswari, J., Parimala, C.: Product sales forecasting and prediction using machine learning algorithm. International Journal of Intelligent Systems and Applications in Engineering (Nov 2023), http://www.ijisae.org/IJISAE/article/view/3801/2435, accessed: 2024-03-12
8. Jordan, M.I., Mitchell, T.: Machine learning: Trends, perspectives, and prospects. Science **349**, 255–260 (2015), https://api.semanticscholar.org/CorpusID:677218
9. Karim, F., Majumdar, S., Darabi, H., Chen, S.: Lstm fully convolutional networks for time series classification. IEEE Access **6**, 1662–1669 (2017), https://api.semanticscholar.org/CorpusID:3708626
10. Lim, B., Zohren, S.: Time series forecasting with deep learning: A survey (2020), https://ar5iv.labs.arxiv.org/html/2004.13408, department of Engineering Science, University of Oxford
11. Lv, Y., Duan, Y., Kang, W., Li, Z.X., Wang, F.: Traffic flow prediction with big data: A deep learning approach. IEEE Transactions on Intelligent Transportation Systems **16**, 865–873 (2015), https://api.semanticscholar.org/CorpusID:7079419
12. Merdas, H.M., Mousa, A.H.: Food sales prediction using mlp, ransac, and bagging. Journal of Techniques **5**(4), 202–208 (2023).https://doi.org/10.51173/jt.v5i4.1458, https://doi.org/10.51173/jt.v5i4.1458
13. Pavlyshenko, B.M.: Machine-learning models for sales time series forecasting. Data **4**, 15 (2018), https://api.semanticscholar.org/CorpusID:53596531
14. Smolic, H.: Revolutionizing sales forecasting: Unleashing the power of machine learning without coding. https://graphite-note.com/machine-learning-sales-forecasting (Sep 2022), accessed: 2024-03-12
15. Supply Chain Brain: Supply chain forecasting: The machine learning revolution (nd), https://www.supplychainbrain.com/blogs/1-think-tank/post/37259-supply-chain-forecasting-the-machine-learning-revolution

16. Weng, T., Liu, W., Xiao, J.: Supply chain sales forecasting based on lightgbm and lstm combination model. Supply Chain Sales Forecasting p. 265 (2019), university of the Chinese Academy of Sciences, Beijing, China; Tianjin University, Tianjin, China. Received 24 March 2019; Revised 23 April 2019; Accepted 10 August 2019
17. Wulf, F.: Neural networks for pattern recognition. Adv. Comput. **37**, 119–166 (1993), https://api.semanticscholar.org/CorpusID:177751
18. Zhang, M., Huang, X.n., Yang, C.b.: A sales forecasting model for the consumer goods with holiday effects. Journal of Risk Analysis and Crisis Responsc (2020), https://api.semanticscholar.org/CorpusID:225616229

Cyber Security

Enhancing IoT Security with Ensemble Deep Learning: A Lightweight, Robust Approach to Intrusion Detection

Huiyao Dong and Igor Kotenko

Abstract The rapid expansion of the Internet of Things (IoT) across various sectors, coupled with the increasing complexity and distribution of information networks, has concurrently introduced intricate security challenges, particularly due to the heterogeneous nature and inherent vulnerabilities of IoT devices. This scenario underscores the urgent need for robust intrusion detection systems (IDSs) capable to protect modern distributed information networks from sophisticated cyber threats. This study introduces a lightweight IDS that can be implemented as a distributed system, leveraging the prowess of deep learning (DL) techniques within an ensemble learning framework to address these challenges. By integrating DL models into ensemble learning frameworks, this approach exploits the collective strength of multiple simple, lightweight models, thereby enhancing the detection capabilities and generalisability of IDSs across distributed IoT environments. Through comprehensive experimental validation on the CICIDS2017 dataset, various ensemble DL strategies, including hard voting, soft voting, stacking, and boosting, are examined. The results demonstrate that this distributed ensemble DL framework significantly enhances IDS performance. In particular, the boosting technique achieved the highest accuracy of 98.31% and a detection rate of 98.32%, while stacking also notably improved the overall performance. This comparative analysis not only underscores the effectiveness of deploying ensemble DL strategies in a distributed IDS for network intrusion detection but also highlights methodological advancements necessary for tackling the complex cybersecurity challenges prevalent in distributed IoT domains.

Keywords Internet of Things (IoT) · Ensemble learning · Deep learning cyberattacks · Intrusion detection

---

H. Dong (✉)
ITMO University, St. Petersburg, Russia
e-mail: hydong@itmo.ru

I. Kotenko
SPC RAS, St. Petersburg, Russia
e-mail: ivkote@comsec.spb.ru

N. Polatidis et al. (eds.), *Intelligent Distributed Computing XVII*, Studies in Computational Intelligence 1203, https://doi.org/10.1007/978-3-031-87639-4_5

1 Introduction

The Internet of Things (IoT) has experienced substantial growth in recent years, revolutionising sectors like smart homes, healthcare, industries, and governmental administration. However, the IoT system is plagued by vulnerabilities, security risks, and privacy challenges. Security issues in IoT have become a major focus, as attacks targeting IoT systems are typically complex due to the diverse nature of devices, their low power capabilities, and the ambiguity surrounding device ownership [7]. The interconnected nodes and devices used for data storage and transmission also expose networks and systems to potential attacks.

To prevent those security threats, intrusion detection systems (IDSs) are the first choice. Signature-based IDS solutions have proven effective against known intrusions but fail against sophisticated ones, while anomaly-based IDS involving machine learning (ML) and deep learning (DL) techniques can filter out unknown network intrusions. The deep, fully connected network, together with a characteristic function, could perform not only anomaly detection but multiple other classification tasks at the same time, indicating its robustness [9]. Apart from deep models, utilising simple binary classifiers like Decision Tree (DT) and K Nearest Neighbours (KNN) with boosting algorithms to build ensemble models is also a feasible solution, which enhanced the weak learners' ability to adapt to complex network traffic classification tasks [3, 4]. In addition to effectiveness, the recent research in the field also emphasises the importance of timeliness and efficiency. Sanju [19] introduced an ensemble framework of recurrent neural networks to improve intrusion detection in IoT systems. This approach used long short-term memory and gated recurrent unit models, along with Harris-Hawk optimisation and fractional derivative mutation for feature selection. Alem et al. [2] proposed an efficient autonomous IDS that combined specification-based and behavioural-based IDS. Privacy concerns were addressed through a federated learning-based anomaly detection mechanism [13]. Lastly, Bertoli et al. [6] utilised unsupervised learning models in an ensemble learning framework for a federated learning-based IDS, demonstrating strong generalisability across heterogeneous networks.

While DL models offer nonlinear flexibility, they may encounter limitations such as high variance, inconsistent predictions, and overfitting resulting from overly complex learned representations. Integrating DL models with ensemble learning frameworks represents a significant advancement. Ensemble learning aggregates the predictions of multiple models to make optimal decisions, thereby improving model performance, generalisation, and relevance in security applications. This strategy helps alleviate overfitting and enhances predictive accuracy on unseen data. Additionally, employing shallow and simple neural networks as weak classifiers within the ensemble DL framework can address individual model limitations, boosting robustness and performance by leveraging diverse architectures. Particularly for complex tasks like network intrusion detection or malware image analysis, which involve high-dimensional data and numerous features, the ensemble DL approach can be practical for learning essential patterns and relationships without necessitating excessively deep and intricate model structures. Building upon the integration of DL models and ensemble learning frameworks described above, it is essential to consider the distinct challenges and requirements posed by designing IDSs for distributed IoT environments. The importance of developing distributed IDS frameworks is paramount, as such a framework offers the advantage of decentralising the detection process and allowing real-time, local intrusion detection at the edge of the network, which aligns perfectly with the distributed nature of IoT networks. The distributed IDS is constructed based on the central monitoring node's big data processing system and hidden agents. Hence, a well-designed ensemble DL approach should be lightweight, adaptive, scalable, and able to function across multiple computing nodes in the IoT environment.

In this study, we conduct a comparative analysis of various ensemble DL learning strategies in the network intrusion domain. We implement three DL models with distinct structures: a fully connected deep neural network (DNN), a convolutional neural network (CNN), and a gated recurrent unit (GRU)-based network. Leveraging these shallow models as weak learners, we deploy and evaluate several ensemble frameworks, including hard voting, soft voting, stacking, and boosting. Our experiments on the CICIDS2017 [20] dataset demonstrate that the ensemble DL approach enhances detection capabilities and generalisability. The boosting algorithm strikes the highest accuracy of 98.31% and detection rate of 98.32%, while stacking can enhance the overall performance regardless of the existence of weak classifiers with under-performance. The lightweight nature of selected DL models ensures that they can be deployed on resource-constrained IoT devices without overwhelming their processing and memory capabilities. Furthermore, the selected ensemble approaches are computationally efficient and allow for the combination of predictions from multiple base models, enhancing the overall accuracy and robustness. By distributing the computation and detection tasks across multiple IoT devices, the system can efficiently process network traffic and detect intrusions in a timely manner. This distributed ensemble learning-based IDS architecture enables scalability, as additional IoT devices can be easily integrated into the system to handle increasing network traffic loads. The system can also adapt to the dynamic nature of IoT environments by continuously learning and updating its models based on incoming network traffic. By leveraging the strengths of ensemble learning and distributed computing, this approach provides a promising solution against evolving cyber threats.

The rest of the paper is as follows: In Sect. 2, we review the current research status in ensemble learning and ensemble DL. Section 3 demonstrates the proposed methodology. Section 4 presents the experiment results of individual weak learners and ensemble models. Lastly, we conclude the paper and discuss the future plan.

2 Related Work

Ensemble learning methods have been extensively studied in the field of network security. A recent survey summarised various ensemble learning methods from both homogeneous and heterogeneous categories [22]. Many studies employing ensemble learning models have focused on binary classification settings. In the research by Fitni et al. [10], ensemble learning models such as Random Forest (RF) and Gradient Boosting (GB) not only surpassed individual ML models like DT and Logistic Regression (LR) but also outperformed the basic DNN model. Similarly, Abrami et al. [1] implemented a stacking classifier using Support Vector Machine (SVM), Naïve Bayes (NB), and RF as base classifiers, with LR as the meta-level classifier; the experiment demonstrated that the ensemble model outperformed all base classifiers. In contrast, Gao et al. [11] reported conflicting results in multi-label classification. While RF and Adaboost achieved competitive overall accuracy compared to a deep DNN model, these ensemble learning models notably lagged behind the DNN in detecting rare attacks. Furthermore, although ensemble learning models generally outperform basic ML models, it is viable to employ simple models for feature selection or importance computation. With the voting ensemble models as the primary intrusion detection technique, using DT to calculate feature importance could result in improved performance [8].

Recently, researchers have realised that the decentralised or distributed IDS can be a feasible solution to the IoT environment, as each node in such a framework can handle data independently and collaborate with other nodes to enhance detection accuracy and threat response time. Moreover, this can reduce dependency on centralised storage and computation systems and enhance the fault tolerance of intrusion detection. Kumar et al. [15] addressed the vulnerability of blockchain to DDoS attacks and presented a novel distributed IDS using fog computing and several models to detect DDoS attacks in blockchain-enabled IoT networks, including RF and an optimised gradient tree boosting system (XGBoost) on distributed fog nodes. Another similar approach is an anomaly detection system decentralising cloud-based security architecture to local fog nodes [14], where RF was chosen as the primary model for achieving the highest accuracy. These works demonstrate the practical value of distributed, ensemble learning-based intrusion detection approaches. Additionally, other ML techniques are viable as well. Sharma et al. [21] proposed a tree-based probabilistic distributed IDS to detect anomalies in service-to-service interactions across multiple edge clusters, specifically designed for the Zero Trust Network environment. Hnamte et al. [12] proposed a distributed detection system utilising a hybrid DL model, incorporating CNN layers, bidirectional LSTM layers, and fully connected layers.

Ensemble DL methodologies present a promising avenue for enhancing network intrusion detection capabilities. Nonetheless, the predominant emphasis remains on optimising the architecture of individual base learners, thereby overlooking the inherent advantages of exploring the ensemble framework itself. While stacking has demonstrated notable success in numerous studies, other ensemble strategies could potentially provide complementary or superior advantages, depending on the context. Thus, there exists a strong rationale for expanding research efforts to encompass a broader array of ensemble techniques, including soft voting and boosting. Consequently, this research endeavours to delve into the examination of diverse ensemble methods. Moreover, a distributed approach to IDSs necessitates efficient models with fewer parameters that can effectively handle the voluminous and diverse data generated by IoT devices. While we examine different parameters to select the appropriate structures of the base models, we aim to ensure that the models are not over-parameterised in order to adapt to the processing capabilities across various nodes.

3 Methodology

3.1 Distributed IDS Design

A distributed IDS comprises a network of multiple IDSs that share logs and alert information. The type and volume of information shared among the IDSs are initially configured together, then fine-tuned adaptively on each node. Real-time network monitoring and detection is independently carried out on each node, while detailed threat analysis is centralised and generated after information centralisation [16]. We propose the distributed IDS as depicted in Fig. 1, reflecting the complexities of a heterogeneous IoT environment, which involves network data processing, distributed monitoring, centralised evaluation, and classification reports, coupled with decentralised response generation strategies.

Given the varied capabilities and protocols among IoT devices, when a traffic packet enters the system, it first arrives at one of the network nodes, equipped to handle diverse network conditions with preliminary filtering applications (e.g., firewalls) that also serve as network traffic load dividers. The packet is subsequently analysed by the local IDS on that node, leveraging decentralised control to enable autonomous decision-making. If identified as malicious by the ensemble DL model, the IDS generates an alert, integrating trust-enhancing mechanisms like blockchain

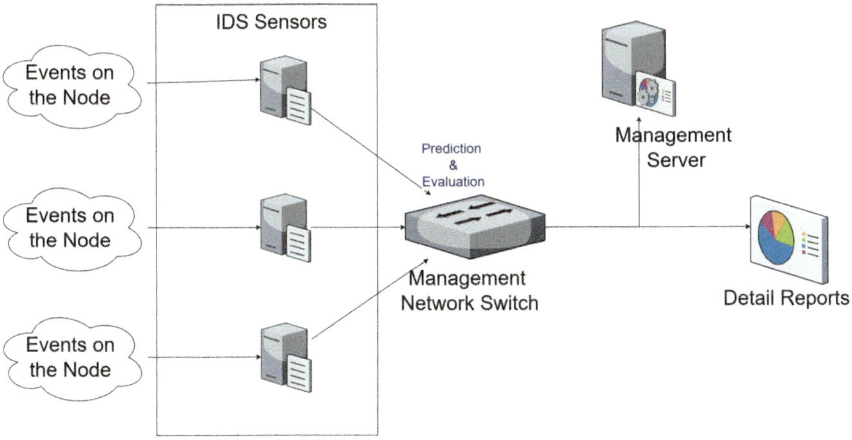

Fig. 1 Distributed intrusion detection system architecture

for secure information exchange. Nodes share information about detected anomalies and updated detection policies, facilitated through a standardised communication protocol to ensure consistent threat detection across the heterogeneous distributed environment. The management network switch, designed for reliability and resilience, aggregates data from all nodes for further centralised analysis and evaluation. Considering vendor and legal constraints, the management server updates detection policies, disseminating them back to the IDS nodes with strategies that respect operational diversity. The effectiveness of this system depends on the detection mode's independent capabilities, standardised protocol for seamless exchange, a robust management network switch, comprehensive reports with detailed traffic analysis, and the timely policy updates and dissemination to nodes, ensuring resilience and trust in response to threats [16].

The remainder of this section will focus on the workflow of data-driven network traffic classification. Figure 2 illustrates the workflow of the proposed technique. Initially, network traffic data is collected from IoT networks and devices, involving the aggregation of traffic traces and session logs. The pre-processing stage includes essential procedures such as data normalisation for uniform data scaling, principal component analysis (PCA) for dimension reduction [18], and the partitioning of data into training and validation sets.

In the data resampling stage, random undersampling is applied to the major classes, namely the benign traffic samples, and the synthetic minority oversampling technique (SMOTE) [5] is employed for oversampling the minor classes. The necessity for resampling arises from the intrinsic class imbalance problem commonly encountered in network traffic data, where benign samples typically far exceed the number of malicious samples. Random undersampling mitigates the imbalanced class distribution by strategically removing certain examples from the majority class. This reduction helps prevent the learning algorithm from becoming biassed towards the majority class, thus allowing it to better focus on the minority class samples during training. The SMOTE-based oversampling technique enhances the volume of minority samples by synthetically generating new examples by examining the properties of KNN in minority class samples and creating new synthetic instances accordingly.

In the model training stage, an ensemble model comprising three base classifiers is implemented: DNN-based, CNN-based, and GRU-based classifiers. By utilising

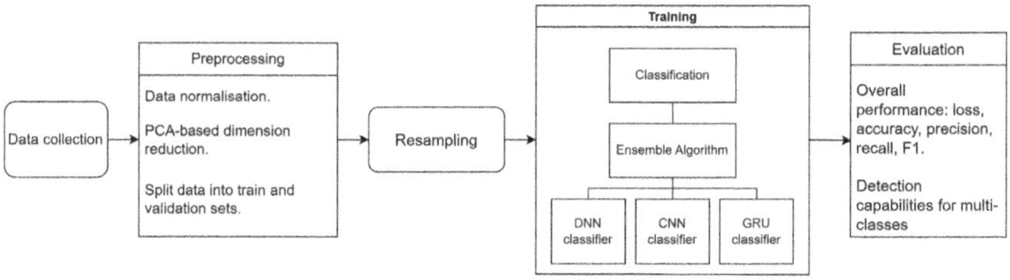

Fig. 2 Workflow of the proposed approach

an ensemble algorithm, the outputs of these classifiers are harmonised to establish a comprehensive classification output that leverages the strengths of each individual model. Finally, the models' overall performance is evaluated using accuracy, precision, recall, and the F1 score, and the detection ability of different network types, especially rare attacks, is investigated. The details of the base classifiers and ensemble algorithms are introduced below.

3.2 Base Classifier

In ensemble learning, the base classifiers independently perform prediction or classification tasks, and their results are combined to make a final prediction. As discussed in [22], the base classifiers should possess diverse learning approaches, accurate individual performance, and the scalability to handle massive data or numerous features. In addition to these characteristics, to optimise the model for adaptation to IoT environments, we aim to implement lightweight DL classifiers that ensure low resource occupancy and high efficiency. We implement three types of DL classifiers, as presented in Fig. 3.

(1) **DNN** serves as the foundational layer in DL models. Within each layer, numerous neurons are employed to capture abstracted features from the received data, along with non-linear activation functions. DNN classifiers are versatile and capable of handling one-dimensional (1-D) sequential data. In the base classifier, we implement five fully connected layers with neuron units of 128, followed by a dropout layer to avoid overfitting.

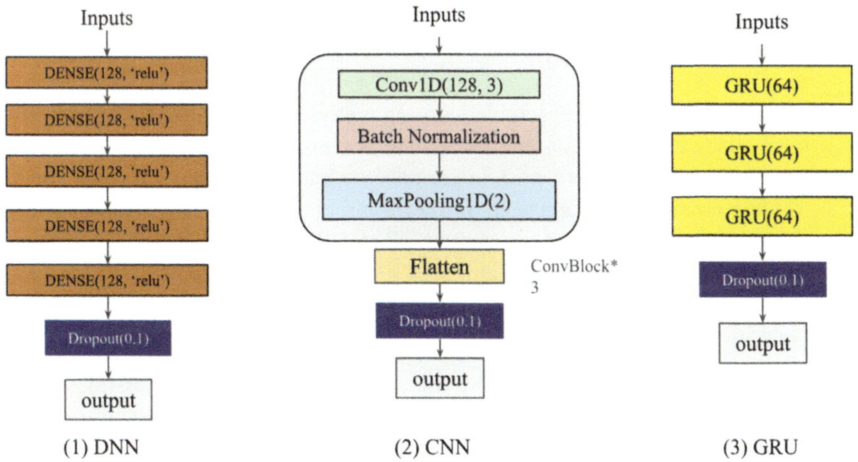

Fig. 3 The structure of the three base classifiers

(2) **CNNs** are specialised in processing grid-like data, such as time series (1-D) or images (2-D or 3-D). The convolutional layers in CNN utilise filters to capture spatial hierarchies. In the classifier, we implement three simple convolutional blocks, each including one convolutional layer with a unit of 128, one normalisation layer to stabilise the data, and a MaxPooling layer for dimension reduction and denoising. Afterwards, a flattening layer and a dropout layer are applied.
(3) **GRU** represents an advanced version of RNN models, incorporating a simplified gate structure to address the vanishing gradient problems associated with the original RNN. Considering the computational complexity of such a layer, this base classifier only contains three GRU layers with units of 64.

Prior to building the ensemble model, we investigated different layer units and model depths. The layer units tested are (64, 96, 128); the implemented Dense layer varies from 1 to 5, while the implemented ConvBlock and GRU layers vary from 1 to 3. Considering the nature of the distributed system, we set upper limits for parameters to ensure the lightweight and computational efficiency of models.

3.3 Ensemble Algorithms

Ensemble learning is a technique that combines multiple models to enhance the stability and predictive performance of the overall model. The ideology behind ensemble learning is to learn several simple models and combine their output to produce a final decision, aiming to combine the strengths of all and minimise the variances and biases of individual models. Assuming any ensemble learning algorithm can be denoted with function G to utilise the prediction of N models, with predicted labels $\{y_1, y_2, \ldots, y_n\}$, the final prediction of the ensemble learning model can be represented by (1):

$$y = G(y_1, y_2, \ldots, y_n) \tag{1}$$

There are two categories of ensemble learning methods: sequential methods, where base learners are generated consecutively, such as stacking and boosting, and parallel methods, where base learners work in parallel, such as voting. Stacking and voting can enhance prediction accuracy via further training and simple aggregation rules, respectively, whereas boosting can decrease bias [17]. Among them, the advantage of voting in ensemble methods is that it uses a simple aggregation rule but can make the final prediction more robust and accurate. Efficiency is the main reason for us to select and investigate this approach.

Voting In a parallel training setting, individual classifiers may develop distinct representations and assign varying labels. To enhance the overall classifier performance, a voting mechanism is employed, which combines predictions either by selecting the class with the highest number of votes (**hard voting**) or by averaging the weighted probabilities of all classifiers' predictions (**soft voting**). Assuming we have three classifiers: $C1$, $C2$, and $C3$. Each classifier predicts the class label for a given input sample. Denoting the predicted labels as y_1, y_2, and y_3, respectively, the final predicted class label using hard voting can be written as (2):

$$y_{\text{final}} = \arg\max_{c} \sum_{i=1to3} I(y_i == c) \tag{2}$$

where the function I represents a Boolean indicator function as follows, with c as the corresponding true labels:

$$I(y_i = c) = \begin{cases} 1 & \text{if True} \\ 0 & \text{if False} \end{cases} \tag{3}$$

For each prediction, we sum up the indicator function values across all classifiers, and the predicted label that can maximise this sum is chosen as the final predicted class label. In soft voting, we consider the predicted probabilities computed by individual classifiers. Denoting the predicted probabilities of class label c by classifier C_i as $p_i(c)$, w_i as corresponding weight, the final prediction by soft voting algorithm is as follows:

$$y_{\text{final}} = \arg\max_{c} \sum_{i=1to3} w_i \times p_i(c) \tag{4}$$

where the predicted probabilities are summed up with corresponding weights. The predicted label that can maximise this sum is chosen as the final prediction. In our experiment, the weights for three classifiers are: DNN-0.3, CNN-0.4, and GRU-0.3, respectively, due to their individual performance.

Stacking Stacking, an ensemble technique, goes beyond simple aggregation by training an additional model to blend and refine the predictions from constituent classifiers. This final predictor, functioning as both a blender and a classifier, takes the ensemble's predictions as inputs to learn a composite representation and generate the final decision. Also referred to as "stacked generalisation," this method employs a more sophisticated strategy by integrating all base classifier outputs into a new model designed to optimise the combination of their predictions. We implement a deep DNN model as this final predictor, and the implementation and training details are presented in Algorithm 1. Initially, different base models (dnn_base, cnn_base, gru_base) are pre-trained individually for 30 epochs with the original datasets to learn prior knowledge. After training the base models, predictions are made on the training and validation data using each of the base models separately. These predictions are then horizontally stacked to create the input features, both for training and validation, for the stacking model. The stacking model contains DNN layers with 64 units, the GeLU activation function, a Dropout layer with a rate of 0.5, and the classification output layer with SoftMax activation. The stacking model is then compiled and trained for 50 epochs.

Algorithm 1 Stacking_Implementation

1: Define base classifier $models = [DNN, CNN, GRU]$
2: **for** $model \in models$ **do**
3: Pre-train the base classifier $model.fit(X, y)$
4: **end for**
5: $X\_stacked$ = horizontal_stacking$[prediction_{DNN}, prediction_{CNN}, prediction_{GRU}]$
6: Define the final predictor $stacking\_model$ with input shape of $n(labels) * 3$ and output shape of $n(labels)$
7: Compile and train $stacking\_model$

Boosting Boosting is an ensemble technique that combines multiple weak learners into a single, stronger model by sequentially training them to refine each other's predictions. This study employs the Adaptive Boosting (AdaBoost) algorithm, as introduced in [23]. The algorithm's key strategy is to concentrate on under-fitted instances, directing subsequent models to address those challenging cases. Misclassified instances are assigned higher weights during training, which are then used to update the weights for subsequent classifiers. In the boosting algorithm, only one base model should be used as the weak learner, and the success of boosting is significantly influenced by this choice. Given that CNN classifiers outperform others in our base models, we opt for CNN as the weak learner for AdaBoost. The CNN is initially pre-trained for 30 epochs, followed by the implementation of ten CNN-based weak learners in the AdaBoost process. The multi-class exponential loss function is employed during training.

4 Experiments

4.1 Dataset

We utilise the public real-world dataset CICIDS2017 [20], which can effectively evaluate intrusion detection and network traffic classification techniques. It comprises weekly users' behaviours across various protocols, including HTTP, HTTPS, FTP, SSH, and email. Table 1 presents the traffic types, distribution of the original, and after applying resampling techniques. Rare but severe attacks like web attack, heartbleed and bot, only have an tiny share originally but are obviously enhanced after the resampling process.

The inherent imbalance in network traffic data can significantly skew the performance of learning algorithms, causing them to exhibit a bias towards the majority class and inadequately detect the minority class-typically the primary focus in intrusion detection contexts. Following the resampling procedure, all classes are balanced.

Table 1 Data summary of CICIDS2017

Traffic type	Original share (%)	Share after resampling (%)
BENIGN	84.3600	12.5
Bot	0.0800	12.5
Brute Force	0.5400	12.5
DDoS	5.0300	12.5
DoS	9.9000	12.5
Heartbleed	0.0004	12.5
Infiltration	0.0014	12.5
Web Attack	0.0900	12.5

4.2 Experiment Setup

The dataset is split into 80% training and 20% validation, with stratification on target labels so that all types of traffic classes are evenly distributed into subsets. To assess the superior performance of ensemble learning compared to DL models, we performed separate experiments with each base classifier for a comparative analysis. All models employed the Adam optimiser with a learning rate of 0.0001 and the categorical cross-entropy loss function. They were trained for 50 epochs and a batch size of 1024, except for stacking and boosting algorithms, where the base models were pre-trained for an additional 30 epochs prior to the main training process.

4.3 Experiment Results

Table 2 exhibits the performance of individual classifiers and ensemble models built upon them. Among the baseline models, the CNN classifier stands out on both training and validation sets; specifically, the accuracy of 98.05% and the F1 score of 98.03% demonstrates its superior performance. In contrast, the GRU model exhibits the weakest performance, with a validation accuracy of 93.88% and a validation F1-score

Table 2 Overall multi-classification performance of CICIDS2017

	Model	Training(%)				Validation(%)			
Baseline		Accuracy	Precision	Recall	F1	Accuracy	Precision	Recall	F1
	DNN	98.97	98.99	98.95	98.97	96.76	96.82	96.72	96.77
	CNN	99.19	99.20	99.18	99.19	98.05	98.08	97.98	98.03
	GRU	97.75	97.88	97.65	97.76	93.88	94.14	93.68	93.91
Ensemble	Hard-voting	99.16	99.17	99.16	99.16	97.41	98.63	97.41	97.92
	Soft-voting	99.18	99.20	99.18	99.18	97.49	98.65	97.49	97.98
	Stacking	**99.35**	**99.35**	**99.34**	**99.35**	98.29	98.30	98.29	98.29
	Boosting	99.24	99.24	99.24	99.24	**98.31**	**98.32**	**98.32**	**98.32**

of 93.91%. A noteworthy observation is that the CNN classifier exhibits minimal differences between training and validation results, suggesting better generalisation, whereas the GRU model displays a tendency towards overfitting.

Ensemble methods generally exhibit enhanced performance compared to individual classifiers, highlighting the benefits of model combinations for improved overall results. Hard-voting and soft-voting surpass DNN and GRU models, but their superiority over CNN is limited due to slightly lower accuracy and F1 scores. This is because voting methods, while leveraging individual models' strengths, may not consistently outperform the best single model when the models possess diverse capabilities, as the aggregation rules are rather simple. The CNN model's excellence in feature extraction and nonlinear modelling provides a substantial advantage over simpler ensemble techniques. Among ensemble methods, boosting demonstrates the best validation performance without signs of overfitting, indicating its effectiveness in enhancing the predictive prowess of the base classifiers. This is partly attributed to the inherent strength of the CNN classifier. Stacking also shows strong performance on the validation set and the best performance on the training set, with an exceptional accuracy of 99.35% and an F1 score of 99.35%, indicating that the meta-learner adeptly combines base models' predictions. Notably, despite GRU's under-performance as an individual classifier, stacking enabled it to achieve high accuracy and F1 scores even when weaker models were included, demonstrating the potential of ensemble algorithms to compensate for weaker base models.

We would like to further investigate the detection capabilities of ensemble learning approaches, particularly their effectiveness with minority classes. The classification reports are presented in Table 3 . Given that the classification of benign behaviours and DoS and DDoS attacks is quite effective, we will focus on those rare attacks with an original share lower than 1%. All models, except boosting, have a recall of 1 for Heartbleed, indicating excellent detection, while the precision is also acceptable. For Brute Force attacks, voting algorithms deliver the best detection as they achieve a recall of 1, whereas stacking and boosting exhibit more balanced performance, with F1 scores of 0.94 and 0.92, respectively. For the other three rare attacks, namely Bot, Infiltration, and Web Attack, although the recall scores are generally high across all ensemble models, the low precision and F1 scores highlight areas for improvement. Nevertheless, the sensitivity to anomalies is crucial for IDS, as the consequence of incorrectly labelling attacks as normal behaviour is more severe than raising a false alarm, which prioritises high recall over low false alarms in this evaluation.

The confusion metrics of all ensemble learning approaches are presented in Fig. 4. Boosting exhibits the lowest false positive rate for corrected benign samples and demonstrates superior performance in detecting major attacks like DoS and DDoS with low false alarms. However, it stands out with the highest false alarms for the infiltration class compared to the other three ensemble methods. This disparity can be attributed to boosting's reliance on a single weak classifier, CNN, in contrast to ensemble algorithms that consider all base classifiers. Notably, despite the Heartbleed attack's limited presence (two cases in the validation set), all models demonstrated effective detection without excessive false alarms. This strong performance on minority classes validates the efficacy of our hybrid undersampling method in addressing

Table 3 Classification report of ensemble classifiers of CICIDS2017

	Stacking			Hard-voting			Soft-voting			Boosting		
Type	precision	recall	F1	precision	recall	F1	precision	recall	F1	precision	recall	F1
BENIGN	1	0.98	0.99	1	0.97	0.98	1	0.97	0.98	1	0.98	0.99
Bot	0.12	0.99	0.21	0.08	1	0.14	0.08	1	0.15	0.12	0.99	0.22
Brute Force	0.88	0.99	0.94	0.86	1	0.93	0.85	1	0.92	0.88	0.97	0.92
DDoS	0.96	1	0.98	0.95	1	0.98	0.95	1	0.98	0.96	1	0.98
DoS	0.95	1	0.97	0.91	1	0.95	0.91	1	0.95	0.95	1	0.97
Heartbleed	1	1	1	1	1	1	0.67	1	0.8	1	0.5	0.67
Infiltration	0.05	0.71	0.08	0.16	0.71	0.26	0.14	0.71	0.23	0.11	0.57	0.18
Web Attack	0.22	0.9	0.36	0.2	0.9	0.32	0.19	0.9	0.32	0.22	0.94	0.36

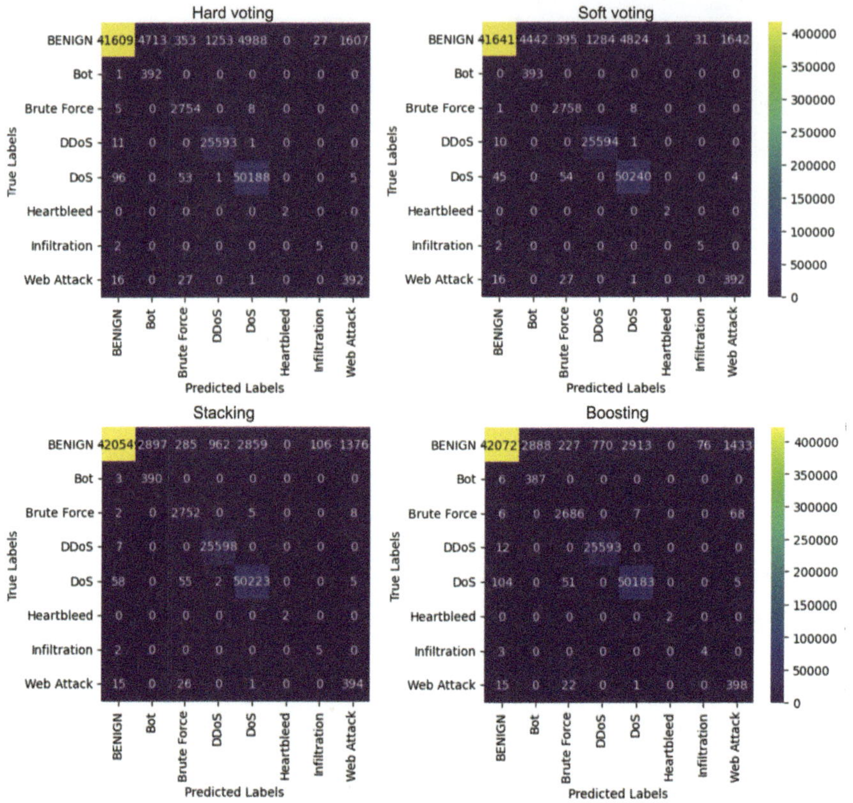

Fig. 4 Confusion metrics of ensemble classifiers

class imbalances within the training data. With misclassified benign cases, the overall false alarm rates for benign behaviours can be calculated. In the test set, the percentage of wrongly classified benign behaviours is as follows: hard voting-3.02%, soft voting-2.94%, stacking-1.98%, and boosting-1.94%. Too many false alarms can significantly affect normal operations; to minimise potential distractions and operational disruptions, stacking and boosting are deemed more appropriate as intrusion detection techniques.

5 Conclusion

As one of the most effective technologies used to enhance IoT security, it is crucial to utilise effective techniques to improve predictive accuracy and generalisation. The successful deployment of distributed IDSs in such environments is essential; however, the detection model should deliver exceptional performance under the constraints of distributed computing resources. Consequently, the corresponding DL model must be expressly designed to ensure both timeliness and accuracy across distributed nodes, even under limited computational capacities.

In this research, we conducted comparative analysis for ensemble DL learning techniques, with both individual classifiers and ensemble algorithms examined. In general, the experiment results suggested that ensemble methods like stacking and boosting can provide significant improvements in classification accuracy and other performance metrics compared to individual classifiers. Furthermore, the adaptability of selected ensemble models facilitates their application in distributed IDS setups, where multiple, lightweight DL models could collaboratively function. This capacity for collaboration without the need for intensive computational resources makes our approach particularly suited for distributed environments.

This study serves as a foundational exploration of ensemble learning techniques for intrusion detection. Our future research will focus on the following directions: (1) Enhancing the model architecture in the stacking algorithm to boost performance and efficiency through thorough experimentation with various model combinations and hyperparameter tuning to pinpoint optimal configurations. Given that the GRU model has shown relatively low performance in this study, it may be advisable to eliminate this model going forward and concentrate on models that do not require GPU resources for enhanced usability. (2) Delving into alternative boosting algorithms within ensemble DL frameworks, building on the success seen in combining Adaboost and deep learning. This investigation will encompass methods such as Gradient Boosting and XGBoost to unveil their potential benefits in improving detection accuracy and speed.

Acknowledgements The study was supported by the grant of the St. Petersburg Science Foundation No. 23-RB-01-09.

Disclosure of Interests. All authors of this paper have no competing interests.

References

1. Abirami, M.S., Yash, U., Singh, S.: Building an ensemble learning based algorithm for improving intrusion detection system. In: Dash, S.S., Lakshmi, C., Das, S., Panigrahi, B.K. (eds.) Artificial Intelligence and Evolutionary Computations in Engineering Systems. pp. 635–649. Singapore (2020)
2. Alem, S., Espes, D., Nana, L., Martin, E., De Lamotte, F.: A novel bi-anomaly-based intrusion detection system approach for industry 4.0. Future Generation Computer Systems (2023)

3. Branitskiy, A., Kotenko, I.: Network attack detection based on combination of neural, immune and neuro-fuzzy classifiers. In: 2015 IEEE 18th International Conference on Computational Science and Engineering. pp. 152–159 (2015)
4. Branitskiy, A., Kotenko, I., Saenko, I.: Applying machine learning and parallel data processing for attack detection in iot. IEEE Transactions on Emerging Topics in Computing **9**(4), 1642–1653 (2020)
5. Chawla, N., Bowyer, K., Hall, L., Kegelmeyer, W.: Smote: Synthetic minority over-sampling technique. J. Artif. Intell. Res. (JAIR) **16**, 321–357 (06 2002). https://doi.org/10.1613/jair.953
6. de Carvalho Bertoli, G., Alves Pereira Junior, L., Saotome, O., dos Santos, A.L.: Generalizing intrusion detection for heterogeneous networks: A stacked-unsupervised federated learning approach. Computers & Security **127**, 103106 (2023).https://doi.org/10.1016/j.cose.2023.103106
7. Desnitsky, V., Kotenko, I., Nogin, S.: Detection of anomalies in data for monitoring of security components in the internet of things. In: 2015 International Conference on Soft Computing and Measurements. pp. 189–192 (2015)
8. ud Din, S.M., Sharma, R., Rizvi, F., Sharma, N.: Detection of botnet in iot network through machine learning based optimized feature importance via ensemble models. International Journal of Information Technology pp. 1–9 (2023), https://api.semanticscholar.org/CorpusID:265299711
9. Dong, H., Kotenko, I.: Multi-task learning for iot traffic classification: A comparative analysis of deep autoencoders. Future Generation Computer Systems **158**, 242–254 (2024)
10. Fitni, Q.R.S., Ramli, K.: Implementation of ensemble learning and feature selection for performance improvements in anomaly-based intrusion detection systems. In: 2020 IEEE International Conference on Industry 4.0, Artificial Intelligence, and Communications Technology (IAICT). pp. 118–124 (2020)https://doi.org/10.1109/IAICT50021.2020.9172014
11. Gao, X., Shan, C., Hu, C., Niu, Z., Liu, Z.: An adaptive ensemble machine learning model for intrusion detection. IEEE Access **7**, 82512–82521 (2019). https://doi.org/10.1109/ACCESS.2019.2923640
12. Hnamte, V., Hussain, J.: Dcnnbilstm: An efficient hybrid deep learning-based intrusion detection system. Telematics and Informatics Reports **10**, 100053 (2023). https://doi.org/10.1016/j.teler.2023.100053
13. Idrissi, M.J., Alami, H., El Mahdaouy, A., El Mekki, A., Oualil, S., Yartaoui, Z., Berrada, I.: Fed-anids: Federated learning for anomaly-based network intrusion detection systems. Expert Systems with Applications **234**, 121000 (2023)https://doi.org/10.1016/j.eswa.2023.121000
14. Kumar, P., Gupta, G.P., Tripathi, R.: Design of anomaly-based intrusion detection system using fog computing for iot network. Automatic Control and Computer Sciences **55**(2), 137–147 (2021)https://doi.org/10.3103/S0146411621020085
15. Kumar, R., Kumar, P., Tripathi, R., Gupta, G.P., Garg, S., Hassan, M.M.: A distributed intrusion detection system to detect ddos attacks in blockchain-enabled iot network. Journal of Parallel and Distributed Computing **164**, 55–68 (2022)https://doi.org/10.1016/j.jpdc.2022.01.030
16. Malani, D., Modi, J., Lilani, S., Desai, Y., Dhanare, R.: Implementation of intrusion detection systems in distributed architecture. In: Cyber Warfare, Security and Space Research. pp. 144–155. Springer International Publishing, Cham (2022)
17. Mohammed, A., Kora, R.: A comprehensive review on ensemble deep learning: Opportunities and challenges. Journal of King Saud University - Computer and Information Sciences **35** (02 2023)https://doi.org/10.1016/j.jksuci.2023.01.014
18. Salih Hasan, B.M., Abdulazeez, A.M.: A review of principal component analysis algorithm for dimensionality reduction. Journal of Soft Computing and Data Mining **2**(1), 20-30 (2021)
19. Sanju, P.: Enhancing intrusion detection in iot systems: A hybrid metaheuristics-deep learning approach with ensemble of recurrent neural networks. Journal of Engineering Research **11**(4), 356–361 (2023)
20. Sharafaldin., I., Habibi Lashkari., A., Ghorbani., A.A.: Toward generating a new intrusion detection dataset and intrusion traffic characterization. In: Proceedings of the 4th International Conference on Information Systems Security and Privacy - ICISSP. pp. 108–116. INSTICC, SciTePress (2018)https://doi.org/10.5220/0006639801080116

21. Sharma, R., Chan, C.A., Leckie, C.: Probabilistic distributed intrusion detection for zero-trust multi-access edge computing. In: NOMS 2023-2023 IEEE/IFIP Network Operations and Management Symposium. pp. 1–9 (2023)https://doi.org/10.1109/NOMS56928.2023.10154326
22. Tama, B.A., Lim, S.: Ensemble learning for intrusion detection systems: A systematic mapping study and cross-benchmark evaluation. Computer Science Review **39**, 100357 (2021)https://doi.org/10.1016/j.cosrev.2020.100357
23. Zhu, J., Rosset, S., Zou, H., Hastie, T.: Multi-class adaboost. Statistics and its interface **2**(3), 349–360 (2009). https://doi.org/10.4310/SII.2009.v2.n3.a8

Clustering-Based Identification of Representative Security Events for Anomaly Detection in Multivariate Sensor Data

Diana Levshun and Igor Kotenko

Abstract Ensuring the security of information systems requires processing many system security events and their correlation. Such events can indicate possible threats, failures or unusual behavior, and their analysis helps to better understand attacks, determine their sources and targets. This is especially important for critical industrial equipment that is constantly monitored by multiple sensors. As a rule, for such systems, it is possible to identify several behavior patterns that are reflected in security events. The paper presents a comparative analysis of methods for identifying representative security events based on clustering of sensor data. These methods allow the grouping of security events without previously known categories, which is often found in real-world data. Each group can be represented by a feature vector that can be used for security tasks. We compare a few clustering methods such as DBSCAN, Mean-Shift and Birch on how well they separate groups of similar events. As an example of the use of representative security events, we solve the problem of anomaly detection in the SWaT (Safe Water Treatment) dataset based on a graph approach and a deep neural network.

Keywords Cybersecurity · Machine learning · Clustering · Event correlation · Anomaly detection · Sensor data

1 Introduction

Security event processing plays an important role in monitoring information systems and industrial equipment. Security events record system states that indicate a possible security breach, either intentional or unintentional. In addition to significant event detection, an important task is to determine the relationship between security events.

D. Levshun (✉) · I. Kotenko
St. Petersburg Federal Research Center of the Russian Academy of Sciences (SRC RAS), Saint-Petersburg, Russia
e-mail: gaifulina@comsec.spb.ru

I. Kotenko
e-mail: ivkote@comsec.spb.ru

N. Polatidis et al. (eds.), *Intelligent Distributed Computing XVII*, Studies in Computational Intelligence 1203, https://doi.org/10.1007/978-3-031-87639-4_6

Both of these tasks are combined into a process called security event correlation, which can identify patterns of behavior and improve overall system security. For example, security event sequence analysis can help better understand the progression of an attack and determine its source and target. Anomaly detection, in turn, aims to look for deviations from normal behavior that may indicate a security breach.

Among the many approaches to correlating security events, hybrid approaches combining several methods are the most popular [14, 16]. Thus, identifying similar security events helps to group them and reduce the amount of data processed. Subsequently, it is possible to analyze the cause-and-effect relationship between groups of similar events or compare a subset of events with known behavior scenarios. Therefore, identifying security events that are similar to each other is a significant step in the analysis of system states.

This paper is organized as follows. Section 2 provides a comparative overview of security event clustering methods. Section 3 describes the experimental analysis and discussion. Section 4 contains the main conclusions.

2 Security Event Clustering Methods

2.1 *Problem Statement*

Let security events as multidimensional sensor data be represented in the form:

$$E = \{e_1, e_2, \ldots, e_n\}, \quad e_i = \{x_{i1}, x_{i2}, \ldots, x_{im}\},$$

where e_n is security event feature vector; n is the number of all events; x_{ij} is the value of the jth feature for i-th event, $i \in [1, n]$, $j \in [1, m]$; m is the feature dimension or number of sensors.

Let us assume that all events can be divided into a number of groups, i.e. set of clusters Ω, using the function μ and a number of input parameters ω:

$$\Omega = \mu(E, \omega) = \{C_1, C_2, \ldots, C_k\}, \tag{1}$$

where k is the number of clusters.

Then each cluster C_i can be associated with some event feature vector e'_i, in such a way that for any event $e_i \in C_i$ the similarity with e'_i is higher than the similarity with representative events of other clusters:

$$C_i \rightarrow e'_i, \quad C_j \rightarrow e'_j, \quad j \in [1, k], \quad j \neq i,$$
$$s(e'_i, e_i) > s(e'_j, e_i),$$

where s is the similarity measure between events, e' is the representative event.

2.2 Overview

The process of clustering security events consists of dividing a set of events in data in such a way that the events of one subset differ significantly in some properties from the events of all other subsets. There are many methods for cluster data analysis [23], which are also applied to time series [13] and sensor data [10]. At the same time, a small part of them is used for clustering security event data, and will be considered as examples of relevant research. Figure 1 provides an overview of the taxonomy of clustering methods for event identification, which we will describe next.

Partitioning-based methods analyze the similarity of each event in relation to the center of mass of the cluster. Thus, the k-means partitions the event data, using the distance from the event vector to the center of the cluster as a measure of similarity. The k-medoids or PAM (Partition Around Medoids) algorithm uses one of the cluster objects (exemplar) to represent the center of the cluster. The CLARA (Clustering Large Applications) applies k-medoids to sample events instead of all objects, randomly selecting data. The Fuzzy C-Means (FCM) uses fuzzy partitioning, assigning an event to each cluster based on the distance between the cluster center and the event vector.

Typically, clustering is used to categorize unlabeled security data and is then complemented by classification methods. Thus, for intrusion detection, Liu et al. [18] combine k-means with a random forest (RF) classifier. Meryem and Ouahidi [21] in turn use a number of classifiers, such as k-nearest neighbors (KNN), naïve Bayes (NB), support vector machine (SVM) and logistic regression (LR). Gadal et al. [8] propose a method for detecting anomalies in network data, which uses the k-means in conjunction with sequential minimal optimization (SMO). Lei et al. [15] use k-means to categorize Android application calls, which subsequently form feature event cluster vectors. These vectors, in turn, are input to a neural network to detect malware calls. Li et al. [17] use FCM in their anomaly detection approach. Clustering is used to generate subsets of multivariate time series, and an anomaly score is then calculated based on measuring the correspondence of the subsets to different clusters.

Density-based methods represent event clusters as high-density regions of data feature space that are separated by low-density regions. The most famous method in this category, DBSCAN (Density-Based Spatial Clustering of Applications with

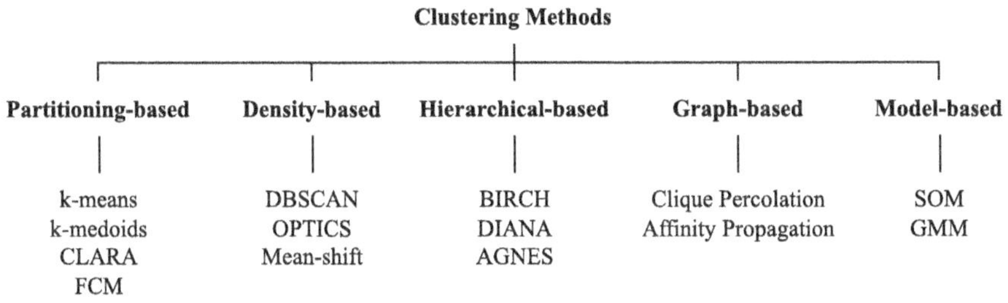

Fig. 1 Overview of clustering taxonomy for RSE identification

Noise), groups events based on the Euclidean distance between them and the minimum number of events in the cluster. OPTICS (Ordering Points to Identify the Cluster Structure) clustering method keeps the cluster hierarchy for a variable neighborhood radius. The Mean-Shift method basically assigns events to clusters iteratively, shifting points in the direction of the highest density.

These clustering methods can be applied as follows. Alfoudi et al. [2] use DBSCAN to categorize security event data into attack types. Garg et al. [9] use k-medoids and DBSCAN together for anomaly detection. Bandara et al. [3] include k-means, PAM and DBSCAN clustering methods to determine optimal time series subsets. For each cluster, the authors then train a LSTM model to predict future values. Dharini et al. [7] propose an intrusion detection mechanism in wireless sensor networks using OPTICS clustering to identify outliers. Markiewicz and Sgandurra [19] use OPTICS to detect unusual behavior of Internet of Things (IoT) devices. Tang et al. [28] improve the Mean-Shift method by replacing Euclidean distance with weighted Euclidean distance as network traffic similarity in attack detection. Niranjana et al. [24] carry out cluster analysis of darknet network traffic using the Mean-Shift to compile attack patterns.

Hierarchical-based methods perform sequential decomposition of a set of events. Depending on the direction of building the hierarchy, researchers distinguish between divisive (top-down) and agglomerative (bottom-up) methods. Common hierarchical methods are AGNES (AGglomerative NESting), DIANA (DIvise ANAlysis), and BIRCH (Balanced Iterative Reducing and Clustering Using Hierarchies), an agglomerative hierarchical clustering method especially suitable for very large datasets.

Examples of studies using this category of methods are the following. Wang et al. [30] propose an approach to detect network traffic anomalies based on the BIRCH for pre-categorization of datasets and an autoencoder (AE) for threshold-based outlier detection. In intrusion detection system by Yin et al. [31] BIRCH and k-means first group the unlabeled data, and then the resulting event label is used as an additional feature for classification with a multi-layer perceptron (MLP). Similarly, Aamir and Zaidi [1] use AGNES together with k-means to label traffic data, which is classified based on kNN, SVM and RF. Naik et al. [22] propose an approach for ransomware detection based on a combination of fuzzy hashing and clustering. At the same time, they conduct a comparative analysis of the k-means, PAM, CLARA, AGNES and DIANA, and conclude that hierarchical-based methods are superior to partitioning-based methods.

Graph-based methods divide the space of event into a finite number of nodes that form a network structure within which clustering operations are carried out. Examples are the clique percolation method (CPM) for fuzzy clustering of an unweighted directed graph and the graph-based affinity propagation (AP) method, which treats all samples in a dataset as possible cluster centers and transfers information through iterations between points of data.

These clustering methods can be used to analyze security event graphs in the following ways. Haas and Fischer [11] use clique percolation to extract loosely connected clusters that potentially contain alerts from a graph of communications between alerts. Studiawan et al. [27] apply CPM to security system log graphs. Tao et al. [29] use an affinity propagation technique to partition alerts into attack scenarios. Raja Sree and Saira Bhanu [25], as well as Chen J. et al. [5], in turn, use AP to detect anomaly behavior in the network. Deng et al. [6] complement affinity propagation clustering with an ant colony optimization, which allows them to efficiently divide the dataset into clusters and then build an anomaly index for each cluster.

Model-based methods assume that there is some mathematical model of a cluster of events in feature space, and seek to maximize the similarity of this model and the available event data. An example of such a method is the self-organizing map (SOM), which is a neural network model. So Hormann et al. [12] use SOM to aggregate security events into attack scenarios. Also popular in clustering problems are Gaussian Mixture Models (GMM), which perform clustering based on the probability distribution for each category, represented as a mixture of several Gaussian distributions. Samunnisa et al. [26] and Bozyiğit et al. [4] complements k-means with GMM to cluster events in intrusion detection.

2.3 *Theoretical Comparison*

Table 1 provides a comparison of reviewed clustering methods for RSE identification. We present a comparison on the following points:

- **time complexity**, where n—the number of all points (examples), k—the number of clusters, n_r—the size of the random sample and p is the number of iterations;
- **interpretability** of the obtained clustering results into the categories "high", "medium" and "low";
- ability to process **noise data** into the categories "high", "medium" and "low";
- **input** parameters, where k—the number of clusters, ϵ—the maximum distance between two points for one to be considered as in the neighborhood of the other, $minPts$—the minimum number of points required to form a dense region, λ—subcluster radius threshold, b—the maximum number of subclusters in each node; l—the size of clique, h and w—the lattice dimensions, c—the number of mixture components;
- **RSE features** for the resulting clusters.

Figure 2 illustrates examples of how RSE features can be extracted from the resulting clusters. The grey dots represent events, reflected in two-dimensional space for convenience. Colored areas indicate clusters, and red plus marks indicate points that may be a representative event for a given cluster. So for k-means this is the mean of points in the cluster, for DBSCAN—the coordinates of the highest density of points in the cluster, for hierarchical-based clustering—the mean of points in subclusters, for graph-based clustering—a representative exemplar of the input events.

The best clustering method depends on the specific characteristics of the data and the goals of the analysis. For practical comparison, we choose DBSCAN, BIRCH

Table 1 Comparative analysis of clustering methods

Method	Time complexity	Interpretability	Noise data	Input	RSE features	References
k-means	$O(nk)$	High	Low	k	Mean of points in cluster	[8, 15, 18, 21, 22]
k-medoids	$O(k(n-k)^2)$	High	Medium	k	Exemplar from data	[3, 9, 22]
CLARA	$O(kn_r^2 + k(n-k))$	High	Medium	k	Exemplar from data	[22]
FCM	$O(n)$	High	Medium	k	Weighted mean of points in cluster	[17]
DBSCAN	$O(nlog_n)$	Medium	Medium	$\epsilon, minPts$	Coordinates of the highest density	[2, 3, 9]
OPTICS	$O(n^2)$	Medium	High	$\epsilon, minPts$	Coordinates of the highest density	[7, 19]
Mean-shift	$O(n^2)$	Medium	Medium	–	Mean of points in cluster	[24, 28]
BIRCH	$O(n)$	High	High	k, λ, b	Mean of points in subcluster	[30, 31]
DIANA	$O(n^3)$	High	Medium	k	Mean of points in subcluster	[22]
AGNES	$O(n^3)$	High	Medium	k	Mean of points in subcluster	[1]
CPM	Polynomial	Medium	Low	l	Exemplar from data	[11, 27]
AP	$O(n^2p)$	Medium	Medium	–	Exemplar from data	[5, 6, 29]
SOM	$O(n^2p)$	Low	Medium	h, w	Mean of points in cluster	[12]
GMM	$O(nkp)$	Low	Medium	c	Probability distributions for cluster	[4, 26]

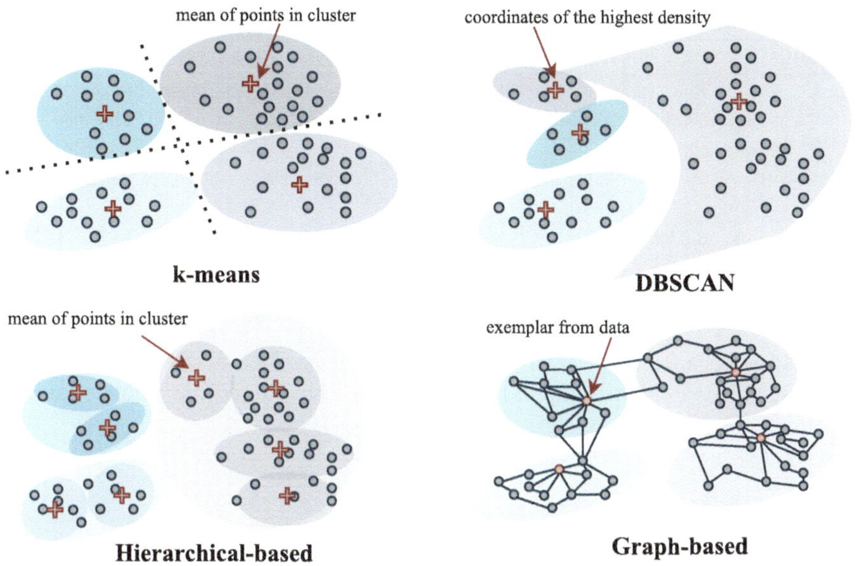

Fig. 2 Examples of RSE identification based on clustering methods

and Mean-Shift as the baselines. These methods have shown themselves to work well in the tasks of clustering security events, including for big data and noisy data. Besides, these methods do not require the number of clusters as an input parameter, which is an important requirement when identifying RSEs. Affinity propagation is another effective clustering method, but its time complexity can be a limiting factor, especially when dealing with large datasets.

3 Experimental Evaluation

3.1 Experimental Setup

We implement experiments using a software tool in Python ver. 3.11 and libraries TensorFlow (ver. 2.15), Keras (ver. 3.0.5), Scikit-learn (ver. 1.4) and others. It runs on AMD Ryzen 7 2.70 GHz, 32 GB RAM, Windows 11.

We use SWaT (Safe Water Treatment) dataset [20] for conducting experiments. The complete dataset contains records from 25 sensors and 26 actuators for an industrial water treatment plant. In this case, water treatment is carried out in six stages, and for each process is designated as P1-P6. SWaT data contains both normal behavior and attacks on various processes. In our experiments, we use a sample from this dataset of $100 \cdot 10^3$ normal events, of which $80 \cdot 10^3$ are training events, and $20 \cdot 10^3$ are test events, as well as $20 \cdot 10^3$ test events containing anomalies. We also identify

Table 2 Analyzed SWat dataset processes

Process	P1	P2	P3	P4	P5
Task	Raw water supply and storage	Chemical disinfection	ultrafiltration and backwash	Dechlorination system	Reverse osmosis
Number of features	5	11	9	9	13
Attack events	8.83%	0.98%	6.96%	3.61%	1.95%

RSE and detect anomalies for P1-P5 process separately, since attacks specifically target these processes in the sample. Table 2 contains a brief description of the analyzed processes.

3.2 RSE Identification

Metrics The number and quality of the obtained clusters in the reviewed studies are usually assessed based on metrics such as silhouette coefficient [2, 3], Davies Bouldin Index (DBI) [9, 30], Calinski-Harbasz Index (CHI) [27], Dunn Index (DI) [22], results of the elbow method [21] and others.

We propose to use the Davis-Bouldin Index and the Calinski-Harbasz Index. These indexes are based on the structure of the resulting clusters and do not require true event labels. This way we can evaluate the quality of unsupervised learning. A lower DBI value and a higher CHI value indicate better clustering.

The DBI measures the average similarity between clusters, while also taking into account the average dissimilarity between each cluster and all other clusters:

$$DBI = \frac{1}{k}\sum_{i=1}^{k} \max_{i \neq j} \frac{D_i + D_j}{q_{ij}}, \tag{2}$$

where D_i and D_j are the average distance between each point in clusters C_i and C_j, respectively; q_{ij} is the distance between cluster centroids c_i and c_j; k is the number of clusters.

The CHI, also known as the Variance Ratio Criterion, is calculated as:

$$CBI = \frac{B}{W} \times \frac{n-k}{k-1} = \frac{\sum_{i=1}^{k} N_i \times \|e'_i - c_E\|^2}{\sum_{i=1}^{k}\sum_{j=1}^{N_i} \|e_{ji} - e'_i\|^2} \times \frac{n-k}{k-1}, \tag{3}$$

where B is the sum of between-cluster dispersion; W is the within-cluster dispersion; N_i is the number of events in the cluster C_i; e'_i is the centroid (representative event) of the C_i; c_E is the centroid of the event dataset; $e_{ji} \in C_i$.

Let the results for several clustering results $\{\Omega_1, \ldots, \Omega_z\}$ be compared with each other. Then we have a set of DBI values:

$$DBIS = \{DBI(\Omega_1), \ldots, DBI(\Omega_z)\} = \{DBI_1, \ldots, DBI_z\}$$

and a set of CHI values for each result:

$$CHIS = \{CHI(\Omega_1), \ldots, CHI(\Omega_z)\} = \{CHI_1, \ldots, CHI_z\}.$$

We normalize these values in each set to the interval [0, 1] as:

$$DBIS^* = \{DBI_1^*, \ldots, DBI_z^*\}, \quad CHIS^* = \{CHI_1^*, \ldots, CHI_z^*\},$$

$$DBI_i^* = \frac{DBI_i - DBIS_{min}}{DBIS_{max} - DBIS_{min}}, \quad i \in [1, z], \tag{4}$$

$$CHI_i^* = \frac{CHI_i - CHIS_{min}}{CHIS_{max} - CHIS_{min}}, \quad i \in [1, z], \tag{5}$$

where $DBIS_{min}$ and $CHIS_{min}$ are the minimum values of the sets $DBIS$ and $CHIS$, respectively; $DBIS_{max}$ and $CHIS_{max}$ are the maximum values of the sets $DBIS$ and $CHIS$, respectively.

We obtain the clustering integral score (CIS), which is defined as the average between the scaled CHI and the inverse of the scaled DBI:

$$CIS = \frac{CHI^* + (1 - DBI^*)}{2}. \tag{6}$$

Given a set of such scores $\{CIS_1, \ldots, CIS_z\}$, we can determine the optimal clustering method Ω as:

$$\Omega = \Omega_i, \text{ where } CIS_i = \max\{CIS_1, \ldots, CIS_z\}.$$

In addition to the quality of clustering, we also evaluate the time spent on training (T_t) and obtaining results (T_c) in seconds.

Models As clustering models we use DBSCAN, BIRCH and Mean-Shift, implemented in the Scikit-learn library. For DBSCAN clustering, we set the minimum number of points $minPts$ as 5 (default) and choose the maximum distance ϵ on the interval [0.1, 1.0] with a step of 0.1. For BIRCH clustering, we set branching factor b as 50 (default) and choose threshold λ on the interval [0.1, 1.0] with a step of 0.1. For Mean-Shift clustering, we choose bandwidth bs used in the flat kernel on the interval [0.1, 0.9] with a step of 0.1.

Table 3 Result of RSE identification based on clustering

Process	Model	Input	k	CHI^*	DBI^*	T_t	T_c	CIS
P1	BIRCH	$\lambda = 0.2$	12	**1.000**	**0.060**	**0.253**	**0.004**	**0.970**
	DBSCAN	$\epsilon = 0.4$	7	0.092	0.140	67.322	0.668	0.476
	Mean-Shift	$bs = 0.1$	24	0.983	0.095	68.676	0.008	0.944
P2	BIRCH	$\lambda = 0.5$	4	**1.000**	**0.000**	**0.279**	0.005	**1.000**
	DBSCAN	$\epsilon = 1.0$	3	0.993	0.040	71.424	0.351	0.977
	Mean-Shift	$bs = 0.7$	5	0.769	0.199	98.726	**0.003**	0.785
P3	BIRCH	$\lambda = 0.1$	77	**1.000**	0.207	**0.467**	**0.021**	**0.896**
	DBSCAN	$\epsilon = 0.9$	4	0.522	0.197	64.927	0.275	0.662
	Mean-Shift	$bs = 0.1$	121	0.668	**0.000**	394.831	0.031	0.895
P4	BIRCH	$\lambda = 0.4$	2	**0.999**	**0.000**	**0.267**	**0.003**	**1.000**
	DBSCAN	$\epsilon = 0.3$	2	0.042	0.999	71.771	0.268	0.021
	Mean-Shift	$bs = 0.7$	2	**0.999**	**0.000**	128.454	0.006	**1.000**
P5	BIRCH	$\lambda = 0.5$	3	0.889	0.065	**0.264**	**0.003**	0.912
	DBSCAN	$\epsilon = 0.2$	2	0.000	0.319	78.929	0.278	0.340
	Mean-Shift	$bs = 0.7$	3	**0.999**	**0.050**	471.907	**0.003**	**0.974**

Results Table 3 contains the security event clustering results for the test set. For Swat processes P1-P5, we present the best result for each clustering model. The original values are rounded to 3 floating decimal places. The best results among the given models are highlighted in bold and the second best are underlined.

We can note that in most cases, BIRCH clustering shows the best results (CIS in the range 0.896–1.0), and DBSCAN shows the worst among the considered models (CIS in the range 0.02–0.977). The Mean-Shift clustering results are relatively comparable in quality to the BIRCH results, and superior for the P5 process. At the same time, training BIRCH models is the fastest, while training Mean-Shift models is quite long. BIRCH can efficiently handle large datasets without requiring excessive computational resources. It uses a tree-based data structure to represent the dataset, allowing for fast processing of data points.

3.3 Anomaly Detection

Figure 3 shows the schematic diagram of the proposed approach.

The input data is records from the sensor system, ordered in time. Data preprocessing consists of extracting features from these records and normalizing them to the interval [0, 1]. We then identify RSEs using the selected clustering method. For the

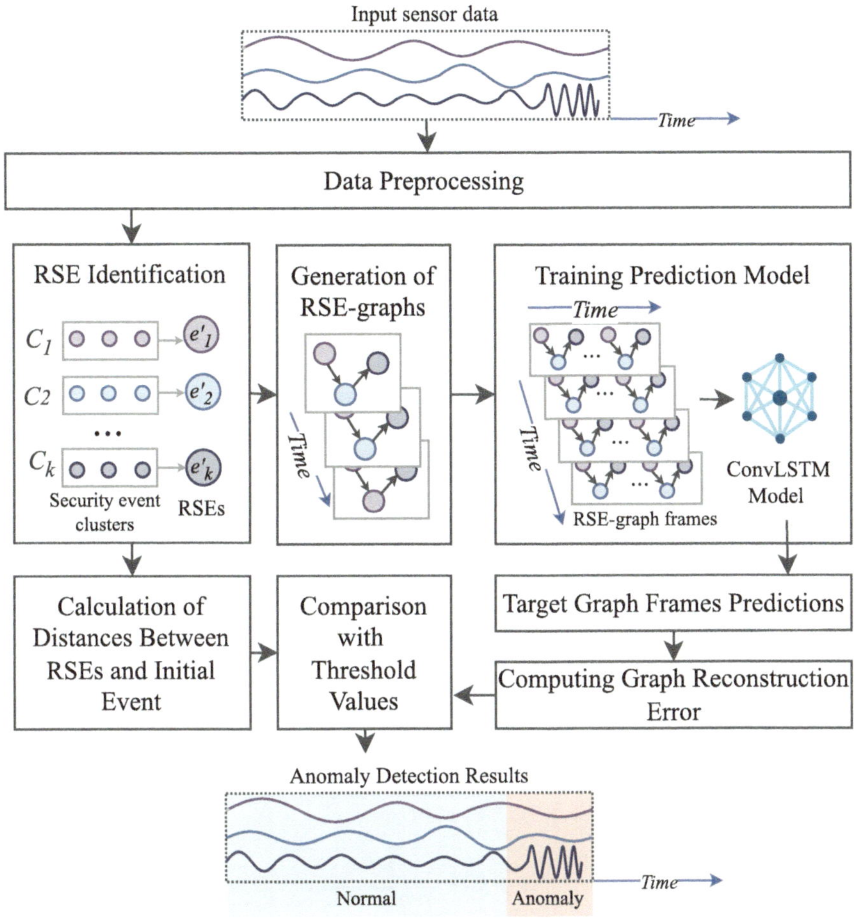

Fig. 3 Overview of the proposed anomaly detection approach

received represenative security events, we calculate the Euclidean distance between them and the initial events.

Then we build RSE-graphs for time periods of the selected length. Each such graph determines the probabilities of transitions between represenative security events in a given time window.

The RSE-graph can be represented as:

$$G = \langle V, R, f \rangle,$$

where $V = \{e'_1, e'_2, \ldots, e'_k\}$ is a set of representative events in the form of graph nodes; $R = \{r(e'_i, e'_j)\}_{i=1,j=1}^{k,k}$ is a set of possible transitions between representative events in the form of graph edges; $f : r(e'_i, e'_j) \rightarrow p_{ij}$ is a function that maps edges to their weights (transition probability).

The choice of this graph format is based on the concept that monitoring transitions between representative events allows us to capture the behavior of the system. Deviations in transition probabilities between these events may indicate anomalies or unusual patterns.

The resulting graphs are divided into sequences of a certain length, called frames. The convolutional LSTM model is trained to predict which graphs will come next given a frame of past graphs. The ConvLSTM model combines the spatial features obtained from the convolution operation with the temporal features obtained from the LSTM.

We calculate the mean square error between the actual data and the prediction as graph reconstruction error. When detecting anomalies, we take into account both the graph reconstruction error and the distance between RSEs and the original events. These two parameters are thresholded using normal data. We set a threshold for the ConvLSTM prediction error (γ) and a threshold for the distance between the original events and the corresponding RSEs (δ). If any of these thresholds is raised, an anomaly is detected.

In this experiment, we identify the RSE for an SWaT anomalous sample based on clustering methods and compare the anomaly detection results. For each process and each clustering models.

Metrics To assess the quality of the approach, we use accuracy (ACC), precision (P), recall (R), F1-score ($F1$) and false positive rate (FPR) over anomaly label data:

$$ACC = \frac{TP + TN}{TP + TN + FP + FN}, \tag{7}$$

$$FPR = \frac{FP}{FP + TN}, \tag{8}$$

$$P = \frac{TP}{TP + FP}, \tag{9}$$

$$R = \frac{TP}{TP + FN}, \tag{10}$$

$$F1 = \frac{2 \times P \times R}{P + R}, \tag{11}$$

where

- TP—True Positive (the model correctly predicts a positive class),
- TN—True Negative (the model correctly predicts a negative class),
- FP—False Positive (the model incorrectly predicts a positive class when the actual class is negative),

Table 4 Results of anomaly detection on SWaT processes

Process	Model	γ, 10^{-3}	δ	ACC	P	R	$F1$	FPR
P1	Birch	0.25	0.25	**0.966**	<u>0.897</u>	**0.727**	**0.802**	<u>0.009</u>
	DBSCAN	1.50	0.20	0.877	0.397	0.501	0.443	0.082
	Mean-Shift	0.30	0.25	<u>0.965</u>	**0.914**	<u>0.714</u>	**0.802**	**0.007**
P2	Birch	1.50	1.00	**0.998**	**0.928**	**0.910**	**0.919**	**0.001**
	DBSCAN	1.50	0.80	0.834	0.032	0.404	0.059	0.160
	Mean-Shift	0.65	1.00	<u>0.972</u>	<u>0.299</u>	**0.910**	<u>0.450</u>	<u>0.027</u>
P3	Birch	0.25	0.50	**0.991**	**0.932**	<u>0.950</u>	**0.941**	<u>0.006</u>
	DBSCAN	2.00	0.80	0.928	0.665	0.097	0.170	**0.004**
	Mean-Shift	0.10	0.50	<u>0.982</u>	<u>0.832</u>	**0.952**	<u>0.888</u>	0.016
P4	Birch	0.25	3.00	<u>0.987</u>	<u>0.756</u>	**1.000**	<u>0.861</u>	<u>0.013</u>
	DBSCAN	2.00	0.01	0.961	0.000	0.001	0.002	0.000
	Mean-Shift	2.00	4.00	**0.992**	**0.836**	**1.000**	**0.911**	**0.008**
P5	Birch	3.00	4.00	**0.984**	**0.796**	<u>0.811</u>	**0.803**	**0.008**
	DBSCAN	6.50	0.01	0.961	0.000	0.000	0.002	0.000
	Mean-Shift	0.50	4.00	<u>0.980</u>	<u>0.716</u>	**0.813**	<u>0.761</u>	<u>0.013</u>

- FN—False Negative (the model incorrectly predicts a negative class when the actual class is positive).

Models As clustering models we use DBSCAN, BIRCH and Mean-Shift with the input parameters determined in the previous experiment. We set the sliding time window size for RSE-graph building as 60 s. We set the graph frame (sequence of graphs) size to 3.

The ConvLSTM model determines the future state in a RSE-graph. The ConvLSTM model includes ConvLSTM2D, BatchNormalization and Conv3D layers. The kernel size for convolutional layers is equal to the number of clusters (k). For the Keras ConvLSTM2D layer, the input data is a 5D tensor: ConvLSTM: $(n, 3, k, k, 1)$. The model is trained using the Adam optimizer with a learning rate of 0.001 in 30 epochs and using early stopping with patience 3. The loss function is mean square error (MSE).

Results Table 4 contains the anomaly detection results for the anomaly set. The original values are rounded to 3 floating decimal places. The best results among the given models are highlighted in bold and the second best are underlined.

We can note that anomaly detection using BIRCH clustering shows better results ($F1$ in the range 80–94%). The approach combined with a Mean-Shift has comparable quality ($F1$ in the range 45–91%). Anomaly detection using DBSCAN clustering has the lowest results ($F1$ in the range 5.9–44%). Thus, we can highlight BIRCH as the optimal clustering method for our anomaly detection approach on the SWaT dataset.

Other methods, such as k-means, k-medoids, CLARA and FCM, use the number of clusters as input. Selecting such a value can be difficult due to low limitations: the number of clusters can be arbitrarily large. The OPTICS method is closely related to DBSCAN, but at the same time its significant disadvantages are memory cost and execution complexity. Hierarchical clustering using the AGNES and DIANA methods is also very time-consuming. Clustering methods based on graphs and models are both resource-intensive and do not always provide a decent interpretation of the results. These are all reasons why these methods were not included in the current experiment. However, they can also be used to cluster security events under assumptions. For example, if there is information about the number of categories of security events or reduced requirements for data processing time.

Clustering helps us detect anomalies by grouping similar security events based on their characteristics. It eliminates the need for extensive knowledge about sensor operating modes. By identifying clusters of ordinary data, we can detect abnormal events that are not included in any cluster or are outliers. Also, the detection of anomalous transitions between representative security events takes into account the temporal sequence of events. Using RSE-graphs allows us to take into account the step-by-step nature of the threat and connect the individual events of the attack together. Graph-based techniques represent event relationships as nodes and edges, aiding in the detection of suspicious patterns. Overall, utilizing clustering and graph-based anomaly detection techniques can enhance the effectiveness of security monitoring and incident response.

4 Conclusion

In this study, we compared a number of methods for identifying representative security events. We conducted an experimental evaluation of the BIRCH, DBSCAN and Mean-Shift methods on the SWaT sensor dataset. We also evaluated the obtained RSEs in the anomaly detection task. Our proposed approach includes both correlation based on event similarity using cluster analysis, and causal correlation using graph model learning. Deep learning is carried out based on convolutional LSTM. The experimental results demonstrate the effectiveness of the proposed approach, as well as the advantage of using BIRCH clustering.

Future work plans include evaluating our approach on other sensor datasets as well as improving the quality of anomaly detection. We also plan to explore additional deep neural network architectures for analyzing representative security events. This will help enhance the scalability and versatility of our approach, making it applicable to a wide range of use cases.

Acknowledgements The study was supported by the grant of the St. Petersburg Science Foundation No. 23-RB-01-09.

Disclosure of Interests The authors have no competing interests to declare that are relevant to the content of this article.

References

1. Aamir, M., Zaidi, S.M.A.: Clustering based semi-supervised machine learning for ddos attack classification. Journal of King Saud University-Computer and Information Sciences **33**(4), 436–446 (2021)
2. Alfoudi, A.S., Aziz, M.R., Alyasseri, Z.A.A., Alsaeedi, A.H., Nuiaa, R.R., Mohammed, M.A., Abdulkareem, K.H., Jaber, M.M.: Hyper clustering model for dynamic network intrusion detection. IET Communications pp. 1–13 (2022)
3. Bandara, K., Bergmeir, C., Smyl, S.: Forecasting across time series databases using recurrent neural networks on groups of similar series: A clustering approach. Expert systems with applications **140**, 112896 (2020)
4. Bozyiğit, F., Türksever, O., Türksever, O., KILINÇ, D.: Machine learning based security analysis: Alarm generation and threat forecasting. International Journal of Engineering and Innovative Research **2**(1), 47–53 (2020)
5. Chen, J., Guo, S., Li, W., Shen, J., Qiu, X., Shao, S.: Network abnormal behavior detection method based on affinity propagation. In: Artificial Intelligence and Security: 6th International Conference, ICAIS 2020, Hohhot, China, July 17–20, 2020, Proceedings, Part II 6. pp. 582–591. Springer (2020)
6. Deng, L., Wan, L., Guo, J.: Research on security anomaly detection for big data platforms based on quantum optimization clustering. Mathematical Problems in Engineering **2022**, 1–10 (2022)
7. Dharini, N., Sowndharya, J., Sudha, P.: Intrusion detection in wireless sensor networks using optics algorithm. In: 2022 International Conference on Applied Artificial Intelligence and Computing (ICAAIC). pp. 1265–1272. IEEE (2022)
8. Gadal, S., Mokhtar, R., Abdelhaq, M., Alsaqour, R., Ali, E.S., Saeed, R.: Machine learning-based anomaly detection using k-mean array and sequential minimal optimization. Electronics **11**(14), 2158 (2022)
9. Garg, S., Kaur, K., Batra, S., Kaddoum, G., Kumar, N., Boukerche, A.: A multi-stage anomaly detection scheme for augmenting the security in IoT-enabled applications. Future Generation Computer Systems **104**, 105–118 (2020)
10. Guyeux, C., Chrétien, S., Bou Tayeh, G., Demerjian, J., Bahi, J.: Introducing and comparing recent clustering methods for massive data management in the internet of things. Journal of sensor and actuator networks **8**(4), 56 (2019)
11. Haas, S., Fischer, M.: On the alert correlation process for the detection of multi-step attacks and a graph-based realization. ACM SIGAPP Applied Computing Review **19**(1), 5–19 (2019)
12. Hormann, R., Bokelmann, D., Ortmeier, F.: Analysis of security events in industrial networks using self-organizing maps by the example of log4j. In: IoTBDS. pp. 51–60 (2023)
13. Javed, A., Lee, B.S., Rizzo, D.M.: A benchmark study on time series clustering. Machine Learning with Applications **1**, 100001 (2020)
14. Kumar, S.S., Agarwal, S.: Rule based complex event processing for IoT applications: Review, classification and challenges. Expert Systems p. e13597 (2024)
15. Lei, T., Qin, Z., Wang, Z., Li, Q., Ye, D.: EveDroid: Event-aware android malware detection against model degrading for IoT devices. IEEE Internet of Things Journal **6**(4), 6668–6680 (2019)
16. Levshun, D., Kotenko, I.: A survey on artificial intelligence techniques for security event correlation: models, challenges, and opportunities. Artificial Intelligence Review pp. 1–44 (2023)

17. Li, J., Izakian, H., Pedrycz, W., Jamal, I.: Clustering-based anomaly detection in multivariate time series data. Applied Soft Computing **100**, 106919 (2021)
18. Liu, C., Gu, Z., Wang, J.: A hybrid intrusion detection system based on scalable k-means+ random forest and deep learning. Ieee Access **9**, 75729–75740 (2021)
19. Markiewicz, R.P., Sgandurra, D.: Clust-IT: Clustering-based intrusion detection in IoT environments. In: Proceedings of the 15th International Conference on Availability, Reliability and Security. pp. 1–9 (2020)
20. Mathur, A.P., Tippenhauer, N.O.: SWaT: A water treatment testbed for research and training on ics security. In: 2016 international workshop on cyber-physical systems for smart water networks (CySWater). pp. 31–36. IEEE (2016)
21. Meryem, A., Ouahidi, B.E.: Hybrid intrusion detection system using machine learning. Network Security **2020**(5), 8–19 (2020)
22. Naik, N., Jenkins, P., Gillett, J., Mouratidis, H., Naik, K., Song, J.: Lockout-tagout ransomware: A detection method for ransomware using fuzzy hashing and clustering. In: 2019 IEEE symposium series on computational intelligence (SSCI). pp. 641–648. IEEE (2019)
23. Nayyar, A., Puri, V., et al.: Comprehensive analysis & performance comparison of clustering algorithms for big data. Review of Computer Engineering Research **4**(2), 54–80 (2017)
24. Niranjana, R., Kumar, V.A., Sheen, S.: Darknet traffic analysis and classification using numerical AGM and mean shift clustering algorithm. SN Computer Science **1**(1), 16 (2020)
25. Raja Sree, T., Saira Bhanu, S.M.: HAP: detection of http flooding attacks in cloud using diffusion map and affinity propagation clustering. IET Information Security **13**(3), 188–200 (2019)
26. Samunnisa, K., Kumar, G.S.V., Madhavi, K.: Intrusion detection system in distributed cloud computing: Hybrid clustering and classification methods. Measurement: Sensors **25**, 100612 (2023)
27. Studiawan, H., Payne, C., Sohel, F.: Automatic graph-based clustering for security logs. In: Advanced Information Networking and Applications: Proceedings of the 33rd International Conference on Advanced Information Networking and Applications (AINA-2019) 33. pp. 914–926. Springer (2020)
28. Tang, D., Man, J., Tang, L., Feng, Y., Yang, Q.: WEDMS: An advanced mean shift clustering algorithm for LDoS attacks detection. Ad Hoc Networks **102**, 102145 (2020)
29. Tao, X.l., Shi, L., Zhao, F., Lu, S., Peng, Y.: A hybrid alarm association method based on AP clustering and causality. Wireless Communications and Mobile Computing **2021**, 1–10 (2021)
30. Wang, D., Nie, M., Chen, D.: BAE: Anomaly detection algorithm based on clustering and autoencoder. Mathematics **11**(15), 3398 (2023)
31. Yin, Y., Jang-Jaccard, J., Sabrina, F., Kwak, J.: Improving multilayer-perceptron (MLP)-based network anomaly detection with birch clustering on cicids-2017 dataset. In: 2023 26th International Conference on Computer Supported Cooperative Work in Design (CSCWD). pp. 423–431. IEEE (2023)

Securing End-to-End Encrypted File Sharing Services with the Messaging Layer Security Protocol

Roland Helmich, Lars Braubach, Theresa Schulz, and Kai Jander

Abstract To protect data on the servers of cloud service providers, file-sharing services rely on End-to-End Encryption (E2EE). However, existing solutions have weaknesses that allow attackers to bypass E2EE permanently after stealing a clients keys once. In this paper, a concept for an E2EE file-sharing service is proposed which does not have this vulnerability. It is based on Messaging Layer Security (MLS) groups for key distribution, an authentication system based on asymmetric cryptography, Attribute-Based Access Control (ABAC) based access rights and a tamper-proof versioned storage system for synchronising sensitive data. The applicability of the concept is demonstrated by a prototype implementation and an evaluation based on benchmarks and a security analysis. Overall, the concept can fulfil the requirements of a basic file sharing service while providing stronger security guarantees than existing solutions.

Keywords End-to-end encryption · Messaging layer security · Data protection · Passkeys · Tamper protection

R. Helmich (✉) · L. Braubach
Bremen City University of Applied Sciences, Bremen, Germany
e-mail: helmich.roland@gmail.com

L. Braubach
e-mail: lars.braubach@hs-bremen.de

T. Schulz · K. Jander
Brandenburg University of Applied Sciences, Brandenburg an der Havel, Germany
e-mail: schuther@th-brandenburg.de

K. Jander
e-mail: jander@th-brandenburg.de

N. Polatidis et al. (eds.), *Intelligent Distributed Computing XVII*, Studies in Computational Intelligence 1203, https://doi.org/10.1007/978-3-031-87639-4_7

1 Introduction

The use of cloud services for file sharing and collaborative working has become the norm for many users. They provide benefits like accessibility of the same data from multiple devices and easy sharing of said data with others. However, one issue with using cloud services is privacy, as the user's data is transferred to, processed and stored on the provider's servers. This paper focuses specifically on the following threats:

1. The data is recorded during transmission between the client and the server. Because many third-party systems are passed through between the client and the server, it is almost impossible to eliminate the possibility of data being intercepted by unauthorised parties.
2. The data stored in the cloud systems is accessed. This scenario can occur not only if the cloud service provider itself acts maliciously, but also through external attacks.
3. A device with access to the data is compromised. This can be the result of a random malware infection or a targeted attack.

To protect against these threats, many services offer E2EE for user data. However not all implementations of E2EE for file sharing services provide the same security guarantees. Two strong security guarantees which are not trivial to achieve are Forward Secrecy (FS) and Post-Compromise Security (PCS). FS describes the property that data sent in the past cannot be decrypted even if the master keys used by participants become known in the future [11]. PCS describes the property that future communication with a compromised participant can be secure again once the compromise has been rectified (e.g. by the removal of malware) [5]. Sometimes PCS is also referred to as self-healing encryption.

This paper proposes a possible solution to provide these strong security guarantees for E2EE file sharing services by using the MLS protocol as a key exchange mechanism combined with a tamper-resistant, versioned storage system for access rules and authentication based on passkeys and signature keys. The results presented here are based on prior work of one of the authors [7].

The paper is structured as follows: Sect. 2 presents a brief review of related work, highlighting the challenges of modern cloud file sharing services. Section 3 introduces the MLS protocol, passkeys and ABAC which forms the foundation of the proposed security approach. In Sect. 4, the novel concept of the E2EE file sharing service is presented. The evaluation and implementation of this concept are detailed in Sect. 5. Finally, Sect. 6 provides a conclusion and an outlook on future work.

2 Related Work

Nextcloud is a file sharing service that offers an E2EE solution, but does not offer FS or PCS. The implementation is described in the *Nextcloud end-to-end encryption* Request for Comments (RFC).[1] It relies on a combination of asymmetric and symmetric encryption to allow users to share E2EE encrypted files. Each user generates an asymmetric key pair for their account and shares the public key with the server and all other participants. Files are encrypted on device using a random key for each file. Whenever someone wants to share an encrypted file, they can encrypt the file's key against the receiving user's public key. The actual implementation uses a metadata file containing the keys of all files in a folder for efficiency, but this has no effect on the security provided. Nextcloud's E2EE is able to secure data in such a way that it can never be seen by the server. It is not possible to intercept plaintext data in transit, even if no additional transport encryption is used. However, no FS properties are provided. If a user's private key is compromised, all data exchanged with that user since the creation of the key can be decrypted. PCS is also not possible with this implementation of E2EE. In the paper *On Post-Compromise Security* by Katriel Cohn-Gordon, Cas Cremers and Luke Garratt, it was shown that PCS can only be achieved by protocols that are stateful [5]. In this context, stateful means that a state is derived from past communication sessions and used in future sessions. The Nextcloud implementation only provides simple public key encryption for key exchange, which has no self-healing capability. After a successful attack, all future files shared with a user are potentially decryptable. This is particularly dangerous if an attack goes unnoticed.

OwnCloud uses a similar setup for its E2EE plugin.[2] Instead of one asymmetric key pair per account, it uses one per device and does not use a metadata file, relying instead on direct key sharing. However, the same fundamental problems with Nextcloud's approach apply here as well.

As can be seen in Table 1, existing cloud services have deficiencies with respect to security. In contrast, the concept that will be detailed next, aims to provide E2EE, FS and PCS in an easy to use and efficient approach.

3 Background

3.1 Messaging Layer Security

The MLS protocol is designed to provide strong security guarantees like FS and PCS for asynchronous, message based communication in large groups. It combines the cryptographic ratchet functions known from the double ratchet algorithm with

[1] https://github.com/nextcloud/end_to_end_encryption_rfc.

[2] https://oc.owncloud.com/rs/038-KRL-592/images/Whitepaper_Data_Protection_and_Data_Secrecy_in_ownCloud_EN.pdf.

Table 1 Comparison of the features of existing solutions with the goals of the concept

	Nextcloud	ownCloud	Concept
Features			
Supports E2EE for content	Yes	Yes	Yes
Direct sharing of E2EE content	Yes	Yes	Yes
Sharing of E2EE with groups	No	Yes	Yes
Automatic sharing of E2EE content based on access rights	Per folder	Per folder	Yes
Security guarantees			
Content is E2EE by default	No	No	Yes
A new random key is used for each object	Yes	Yes	Yes
Keys stolen in an attack cannot be used to decrypt previously shared content (FS)	No	No	Yes (for content that is no longer shared)
Keys stolen in an attack cannot be used to decrypt all future shared content (PCS)	No	No	Yes

the principle of Asynchronous Ratcheting Trees (ARTs), first described in the paper *On Ends-to-Ends Encryption: Asynchronous Group Messaging with Strong Security Guarantees* by Katriel Cohn-Gordon et al. [6]. This allows keys to be exchanged efficiently. For a group with n members, the complexity is reduced from $O(n)$ to $O(logn)$. The actual sending of messages is not specified as part of MLS. An external Delivery Service (DS) must be used to deliver the messages to all group members. Group members are authenticated using credentials. Version 1.0 of the protocol supports *basic* credentials and certificate-based *x509* credentials. The MLS protocol expects the use of a Authentication Service (AS) to authenticate the credentials provided by group members. According to the *Messaging Layer Security (MLS) Architecture* [3] document, an AS must perform three tasks:

1. Issuing credentials to members that attest to the connection between their identity and the signature keys used.
2. Enabling the verification of the credentials presented by a member by other members.
3. Allow a group member to verify that a credential represents the same member as another credential.

The algorithms used for encryption and signing are interchangeable in the MLS protocol. Version 1.0 of the MLS protocol was published as RFC 9420 on 18 July 2023 [2]. The security of the MLS protocol has been analysed thoroughly. In [4] the security of the MLS protocol was proven using a formal model. The publication [1] also deals with the security of the protocol as part of a theoretical analysis. The security guarantees claimed by MLS are confirmed by this research.

3.2 Passkeys

Passkeys are a new form of authentication that provide users a way to log in without a password. They are based on the Fast Identity Online (FIDO) Universal Authentication Framework (UAF), which is a FIDO Alliance standard. Passkeys and FIDO UAF are described in the *FIDO UAF Architectural Overview* [10]. Authentication with passkeys is based on the generation of signatures using asymmetric cryptography. A passkey therefore consists of a signature key pair, the private key of which is stored by an authenticator. For each service, referred to as a Relying-Party (RP), the authenticator generates a separate key pair during registration. The public key is transmitted to the RP and associated with the user account. Users can use passkeys to log in by signing a random challenge generated by the RP. Depending on the type of authenticator used, a PIN or biometric features such as a fingerprint may be required to allow the signing process. In addition to authentication, the signatures generated by passkeys can also be used for other purposes like confirming transactions by signing them locally. Manipulation by third parties is then no longer possible without invalidating the signature. To use passkeys, users do not necessarily need a dedicated hardware authenticator, because most operating systems already natively support the functions of an authenticator. These include Android, iOS/iPad OS, macOS, chromeOS and Windows 10/11.[3]

3.3 Attribute Based Access Control

ABAC describes a variant of access control based on a logical basis. The National Institute of Standards and Technology (NIST) defines ABAC in the Special Publication 800–162 *Guide to Attribute Based Access Control (ABAC) Definition and Considerations* [8]. In the context of ABAC, subjects are those who want to access objects and perform certain operations on them. Attributes are key-value pairs that describe properties and can be assigned to both subjects and objects. Rules can than be defined that form a decision based on the attributes of subject and object. By using attributes for both subject and object, complex access policies can be expressed with ABAC. For example, the required clearance level could be attributed to an object

[3] https://passkeys.dev/device-support/.

and the assigned clearance level to a subject. Then, a simple rule that compares the two attributes can be used to handle all clearance levels. Any new clearance levels that may be created in the future would also be compatible.

The NIST regards ABAC as superior to the more traditional Role-Based Access Control (RBAC), in part because of this flexibility. In general, RBAC is described by NIST as a specialised variant of ABAC. This means that all access policies that could be implemented with RBAC can also be implemented with ABAC.

4 Concept for an End to End Encrypted File Sharing Service

The concept proposed in this paper must meet a number of requirements. Firstly, in terms of the security guarantees provided, all files must always be E2EE. Furthermore, the concept must provide PCS and FS for user data. Since users should be able to share existing files with new users, the FS property is interpreted more loosely. It must only apply to files that have been removed from the service by deletion or unsharing. These security guarantees must be maintained without significantly affecting the usability of the service. All the operations that users expect from a file sharing service should still be possible. This includes uploading/downloading files, sharing files, setting access control rules, adding/removing devices from an account, etc.

The basic idea to achieve this is to implement E2EE by allowing clients to exchange keys directly with each other using MLS groups. This means that the strong security guarantees and scalability of the MLS protocol can be relied upon for the key exchange itself. The encryption keys are stored only on the clients. The server does not directly participate in the key exchange and only provides the infrastructure needed for MLS. Access rights are realised via an ABAC system. In order for keys to be shared directly between clients, access rights must be evaluated on the client side. This way, when a file is uploaded or modified, it can be determined locally to which other clients the key needs to be sent. Figure 1 gives a basic overview of what data is stored and where it is stored.

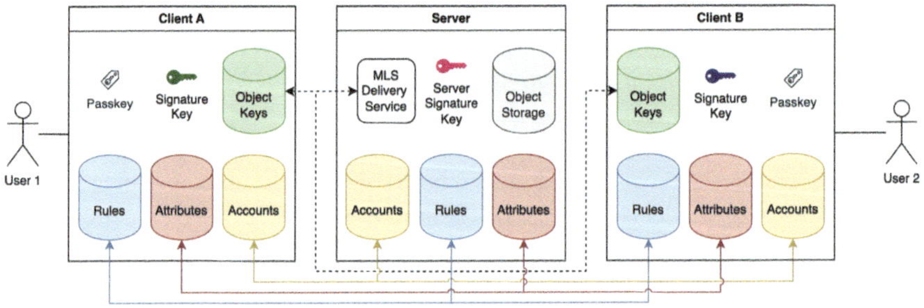

Fig. 1 Basic structure of the concept

4.1 Tamper-Resistant Versioned Storage for Shared State

The security of the concept relies on the fact that the server and all clients have a common view of access rules and attributes as well as all signature keys and passkeys. Additionally, this information may only be modified by authorized users. To ensure this, a versioned storage system based on signed, chained commits is used. Actual user data/objects are not stored in this system.

The versioned storage system is structured as follows: similar to the version control tool Git, data is kept in repositories that consist of commit objects and commits. A commit always references a list of commit objects that represent the state at the time of the commit. Each commit is uniquely identifiable by its hash value and references the hash of its predecessor. All commits are signed by a passkey or device key linked to an account. This results in a chain of commits that represents the history of the storage system. Commit objects are also identified by their hash value. The structure of this system is shown in Fig. 2.

To apply a commit, a client must send it to the server, which then checks whether the commit is valid and whether the client has the necessary rights to apply the changes made by the commit. If this is the case, the commit is applied and thus represents the new state of the storage system. Other clients can then retrieve the new commit, also validate it and, if successful, apply it to their local copy of the repository.

The server always keeps a complete history of all commits and objects. Clients only need to store the latest accepted commit and the commit objects that it references. This type of storage system offers a number of advantages. To ensure that two clients have the same view of a repository, it is sufficient to compare the latest commit hash. At the same time, the linking and signing of commits ensures that the history of the

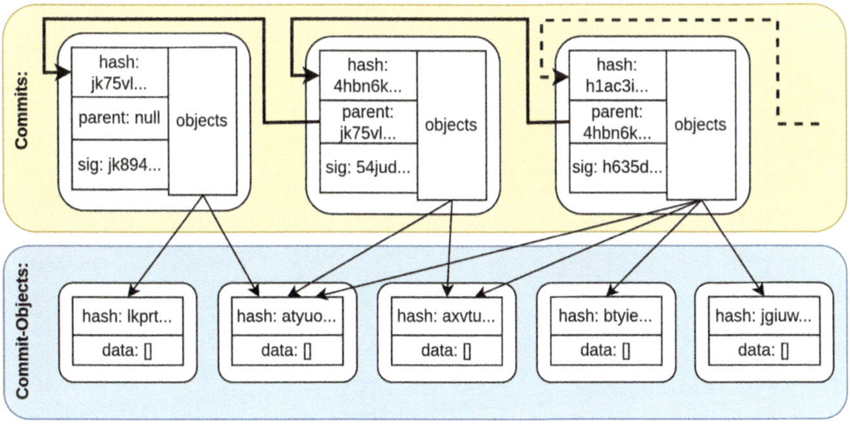

Fig. 2 Example for a chain of signed commits referencing commit objects

storage system cannot be manipulated, making all changes auditable. The biggest disadvantage is that parallelization is not possible due to the chaining of commits.

To ensure that all clients have the same view of the repository, they must synchronize the current commit hashes with other clients at regular intervals. This prevents a compromised server from not delivering commits to some clients.

4.2 Authentication and Signature Keys

Signature keys are required so that users can uniquely prove that commits have their permissions. In addition, MLS also requires the use of a digital signature system.

Each device has its own key pair for signing, the device key. This key pair is used exclusively to generate and validate signatures that occur during normal use of the file sharing service. This includes, for example, signing files before uploading, signing commits or, as required by the protocol, signing messages. The private key must never be shared and remains bound to the end device on which it was created for its entire lifetime. If supported by the hardware, a non-extractable key managed by an Hardware Security Module (HSM) should be used.

In addition, passkeys (FIDO UAF) are used to allow users to log in and authorize particularly security-critical operations. These include:

1. Adding a new device (associates a new device key with the account)
2. Removing a device (invalidates an existing device key)
3. Adding another passkey
4. Removing a passkey.

Commits that contain changes to this data must always be signed by a passkey associated with the account. The passkeys therefore represent the highest instance in the management of a user account. At the same time, the use of a passkey is only rarely necessary for users, so that working with the service is not often interrupted by authentication queries. This results in the hierarchy shown in Fig. 3.

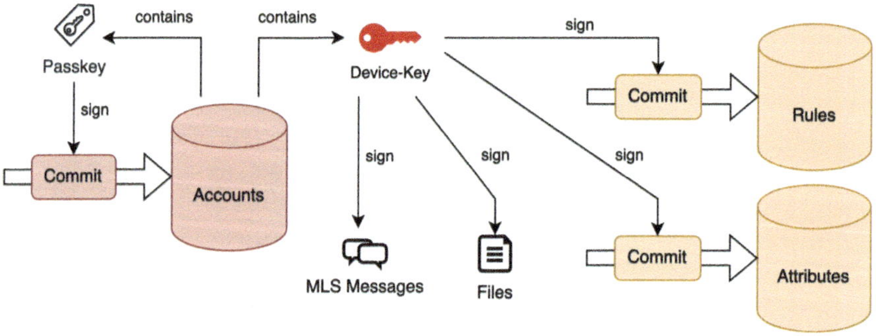

Fig. 3 Hierarchy of signature keys

All clients must know all existing passkeys and signature keys so that they can validate signatures independently of the server. The Trust On First Use (TOFU) principle is used here, meaning that clients trust the first passkey they see for a user account. Signature keys and other passkeys introduced by commits signed by this passkey are also trustworthy. Only when a new passkey is added and then used to invalidate the original one does it lose its validity.

Because TOFU is used, it must be possible for users to validate the keys associated with each user account through independent communication channels. This can be used to ensure that a key actually belongs to the claimed person.

Loss of Passkeys A problem arises if a user loses access to all passkeys associated with an account. There are a number of strategies that could enable access to be restored in such a case. These include, for example, security questions or the use of recovery passwords. In the article *Evaluation of Account Recovery Strategies with FIDO2-based Passwordless Authentication* by Johannes Kunke et al., a comparison of different recovery methods for FIDO UAF based logins was made [9]. The use of a backup token was found to be the best practical method that does not weaken security. The other methods that were rated as good also always include at least one additional token as a backup; they differ mainly in how this backup token is used. Purely knowledge-based recovery methods such as security questions or recovery passwords are not recommended, as their use would cancel out the security benefits gained from passwordless login.

For the file sharing service account recovery should thus not be allowed if all passkeys are lost. However, since passkey-capable authenticators can be purchased inexpensively, it is reasonable to expect users to keep a backup authenticator on hand. In addition, some passkey implementations already offer backup functions. For example, Apple synchronizes passkeys with the iCloud Keychain by default, which significantly reduces the chance of loss.

4.3 Encryption and Key Exchange

Before an object is to be uploaded, it and its name must be encrypted. To do this, the end device must obtain a key from a cryptographically secure random source, which is then used once to encrypt the data using a symmetric method. After encryption, the key is stored in the local key ring of the device. This contains the keys for all objects that a device is authorised to access and should, if possible, be protected with a HSM.

The next step is to distribute the key to all other devices that are also authorised to access the object. This requires an evaluation of the access rights, which is done on device to prevent manipulation by the server. Once the authorised devices have been determined, a suitable MLS group with exactly these participants must be found or created. The key is then sent to this group as a message together with the information required to assign it to the respective object. The group members add the key to their local key ring after receiving the message, which completes its distribution.

In addition to the direct distribution after the creation of new objects, keys may have to be redistributed if the list of authorised devices changes. This can occur when access rights are changed, when a new device is added or when a new user account is created. All keys that are available to the device that initiated the change can be redistributed directly by that device. However, when rules or attributes are changed by administration accounts, it may be the case that these accounts themselves do not have access to all the objects affected. It is therefore also possible for clients to request keys for objects from other clients via MLS groups. If a client determines after a rule or attribute update that it should have access to new objects but has not yet received the keys, it can request all other authorised users to resend the keys via an MLS group. This procedure can lead to situations in which access to objects is delayed because the corresponding keys have not yet been redistributed. How often this occurs depends heavily on the way the service is used.

A client that has been denied access to an object must delete the key for this object. In the special case that a change causes all currently authorised users to lose access, but at the same time new authorised users are defined, the key must be transferred by distributing it via a MLS group. It should not be possible for an object to exist without access authorisation.

5 Evaluation of the Concept

The evaluation of the concept consists of two parts. In the first part, the security of the concept is analysed on a theoretical level. The second part aims to prove applicability by implementing a prototype and running benchmarks.

5.1 Analysis of the Security Architecture

In the following sections an analyse of three different attack scenarios is made. This is not a comprehensive security analysis, but should give a good overview of the capabilities and shortcomings of the concept.

Recording of Data During Transmission The first attack class considered is the recording of data during transmission. The architecture of the Internet makes this type of attack easy to implement. It is generally assumed that transport encryption such as Hypertext Transfer Protocol Secure (HTTPS) is used to protect the communication between client and server. This alone would make it difficult for an attacker to read intercepted data. Assuming that no additional transport encryption is used or that it has been bypassed by an attacker, there is still a high level of protection.

Sensitive contents such as files/objects are always encrypted and the keys are distributed exclusively via encrypted MLS messages. This means that an attacker is still unable to interpret sensitive data. However, it would be possible to reconstruct the contents of the versioned storage system by recording queries. The data it contains is not secret. However, they could be used to determine which devices have access to which objects, and thus presumably the secret keys to those objects. Furthermore, it is conceivable that attributes could contain sensitive data that could be used to infer the identity of users, for example. The extent to which this is a problem depends on the specific application.

Compromise of the Server The next class of attacks involves attacks on the server. For the sake of simplicity, it is assumed here that the worst case has occurred and the server is completely under the control of an attacker.

The concept is designed to resist this type of attack and can therefore still guarantee security in this situation. The server has access to all data/objects, but since these are encrypted and the keys are only stored on the clients, no access to sensitive content is possible. Similarly, the server has access to all commits and commit objects in the versioned storage. It is conceivable that an attacker could attempt to change user rights by manipulating commits and thus gain access to content. However, since all changes must be signed by a authorised client, the server cannot create manipulated commits that would also be accepted by the clients. Access to the MLS Delivery Service also offers an attacker few advantages. As described in RFC 9420, MLS is designed in such a way that a compromised DS can only suppress messages, but never gain access to their content [2]. It is also not possible to send manipulated messages, as these must be signed.

The greatest threat is that the server could withhold new commits or messages. The server could prevent a user from removing a compromised device by not forwarding the corresponding commit to other clients. The concept envisages that clients synchronise the status of the versioned storage with each other at regular intervals via a MLS message. Such a comparison would reveal that the server had withheld commits. However, the state between the clients may be inconsistent without a way of detection until such a comparison message is received.

In addition, the server can try to suppress MLS messages in order to prevent the versioned storage from being synchronised. However, since the server cannot view the message contents, it is not possible to deliberately suppress certain messages. If all messages from and to a client are suppressed, the client is effectively rendered inoperable. If, however, only a few messages are withheld, this can be detected by comparing the generation counter, which is part of all MLS messages [2].

Man-in-the-Middle Attack A security risk can arise if an attacker can impersonate the server to users. If a new account is created and a device is added in this situation, it is theoretically possible to carry out a Man-in-the-Middle (MITM) attack. Since the new device has never communicated with other participants, an attacker can pretend to be a fake server instance to the new device. He can then present a new account under his control to the existing devices. This means that an attacker can intercept all the keys shared with this client and impersonate it as they wish.

The concept provides protection against this type of attack by giving users the option to verify identities in an out-of-band comparison. For example, signature keys are compared in person by displaying them as QR codes and scanning them reciprocally. This makes it possible to detect an attack of this kind. However, a problem with this type of protection is that it only works if users actually check the identities. Since this requires a certain amount of effort, it is not unlikely that this step will be skipped.

Compromise of a Client The last class of attacks to be examined here are attacks on clients. Since the keys are stored locally on the clients, such attacks are particularly dangerous.

If an attacker manages to compromise a client, it is highly likely that they will be able to access all the keys that this client stores locally. This means that all objects currently shared with this client can be viewed. In addition, an attacker can also make commits and sign them on behalf of the client. However, it should be noted that an attacker can only act within the limits of the rights of the account to which the client is assigned. Since commits with a passkey signature are required to change the account, even if the client is completely compromised, an attacker cannot simply add a new device to an account or remove existing devices. To do so, he would also have to get the user to sign a corresponding commit, since the use of passkeys always involves user interaction.

The FS guarantee provided by MLS messages means that even if an attacker has full access to the current MLS group state, it is not possible for him to decrypt messages that were sent before the client was compromised [2]. The concept also requires that keys for objects to which access has been revoked must be deleted by the client. It is therefore not possible for an attacker to obtain the keys to these objects if they did not already have access to the client at the time of the deletion/revocation.

The concept provides for keys to be shared with a client again after a compromise has ended, without an attacker being able to intercept them. MLS offers PCS for a group as soon as a client has executed a key update that could not be observed by an attacker [2]. An attacker can only continue to read messages if he can continue to observe all key updates after the compromise has ended. A single unobserved key update is enough to restore the PCS guarantee. Since such updates are carried out regularly, it is likely that the key exchange will quickly be secure again.

5.2 *Implementation of a Prototype*

To test the basic functionality of the concept, a prototype was developed for both the server and the client. The aim was to realise a minimal implementation for a few selected functions, on the basis of which the concept could be validated and possibly further developed at a later date. The focus was on the following functions:

1. Key exchange via MLS groups
2. Versioned storage of attributes, rules and account data

3. A functional account management system with corresponding commit validation
4. A simple implementation of ABAC based access rights

Other features were not implemented, or were only rudimentary. For example, uploading and downloading files/objects is a problem that has been solved many times before, and it would not add any value to revisit it here.

The server prototype was implemented using the C# programming language with Microsoft's .NET8.0 Software Development Kit (SDK). Instead of a single application, four micro-services were used to implement the functionality:

- The gateway service that represents the interface to the outside world. It must provide a Representational State Transfer (REST)-Application Programming Interface (API) for clients and handle communication with the other services via a message bus.
- The version control service that implements the versioned storage system. It is used to store rules, attributes and account data.
- The account service that is responsible for assigning account IDs and validating commits that affect accounts.
- The MLS Delivery Service that is responsible for distributing MLS messages and KeyPackages.

The client prototype was implemented using the Rust Programming Language. Rust was chosen because OpenMLS provides an implementation of MLS for this language, which implements the protocol as specified in RFC 9420[4]. Version 0.5.0 of the library was used for the prototype. As the client prototype is mainly used for concept validation and benchmarking, no real passkeys have been implemented. They are instead simulated by standard Edwards-curve Digital Signature Algorithm 25519 (Ed25519) signature keys. Since passkeys require user interaction, their use in a benchmark with many simulated clients would be impractical at best.

Benchmarks To test the key exchange, a scenario is used that simulates instances of the file sharing service. An instance with 100 clients and an administrator account and an instance with 1000 clients and an administrator account are considered. For the benchmarks, an increasing number of keys are shared, starting with one and ending with 10,000. The time taken to distribute the keys and the average time taken for each simulated client to receive them is measured. The benchmarks where executed on a 2021 Macbook Pro 16.2 (32 GB RAM, M1 Max) and the results are depicted in Table 2.

A number of interesting observations can be made. For one thing, sharing a single key takes a relatively long time if there is no corresponding MLS group to distribute it to. This is partly because key packages have to be retrieved from the server for each participant in order to initialise a new group. Creating a group is also more cryptographically complex than simply sending messages. For 1000 clients, this process is therefore significantly more time-consuming, even per client. For 100 clients, the initial sharing takes about 6.8ms per client, and for 1000 clients about

[4] https://openmls.tech/.

Table 2 Benchmark results

Action	Number of keys	100 clients	1000 clients
Share key (init MLS group)	1	681.67 ms	15923.25 ms
Receive key (join MLS group)	1	112.21 ms	572.65 ms
Share keys	10	86.05 ms	231.14 ms
Receive keys	10	34.82 ms	66.70 ms
Share keys	100	380.71 ms	755.87 ms
Receive keys	100	101.14 ms	133.21 ms
Share keys	1000	7515.76 ms	7344.77 ms
Receive keys	1000	614.10 ms	646.25 ms
Share keys	10000	61057.49 ms	85289.02 ms
Receive keys	10000	5504.31 ms	5709.48 ms

15.9 ms per client. It is clear that key distribution is much faster after the initial creation of the group. The efficiency of MLS is particularly evident when comparing the two instances. For example, distributing 10,000 keys to 1,000 clients takes only 40% longer than distributing the same number of keys to 100 clients.

6 Discussion

Overall, the concept is able to fulfil the goals set in Table 1. An E2EE file sharing service can be developed on its basis, offering strong security guarantees such as PCS and, with some limitations, FS. Thanks to the use of an ABAC based rights system, it is possible to offer extensive options for access control and content sharing without compromising on security. The prototype implementation also proved that the system can be implemented in practice and used at scale. Since the concept is based on an existing protocol that has been analysed several times for security, it can be assumed with a high degree of confidence that it can actually fulfil the security guarantees.

However, the concept also has weaknesses. Firstly, shared content may, in some circumstances, not be immediately available because no client who has the encryption keys is currently online. There is presently no simple solution to this problem that does not also weaken the security guarantees. Existing solutions do not have this problem because they do not support situations where rights to an object are granted by third parties who do not have access to the content themselves. It is difficult to assess how serious this disadvantage is without testing the system in a real environment.

Another disadvantage of the concept is that the versioned storage is always visible to all clients. This means that any client can see all the attributes assigned to each object and account, all the existing rules, and all the signature keys and device keys associated with any account. This data is not sensitive in itself, as objects are only identified by a random ID and the real name is encrypted. However, depending on the use of attributes or rules, it may be desirable that they are not fully visible to all clients. Further research would therefore be needed in this area.

7 Summary

This paper investigates how strong security guarantees can be provided to E2EE file sharing services using the MLS protocol. Existing E2EE solutions used by Nextcloud and ownCloud have vulnerabilities that allow an attacker to decrypt files shared in the past and in the future if they can gain access to a user's keys just once, for example through a malware infection. A concept for an E2EE file sharing service is presented that addresses these issues and provides stronger security guarantees as well as more functionality. The concept uses the MLS protocol for key distribution and ABAC for access rights, while a versioned storage system similar to Git is used for tamper-resistant storage of attributes, rules and signature keys. For signing and authentication, a combination of passkeys and regular signature keys is used. A number of theoretical attack scenarios are investigated for evaluation. A prototype is built and benchmarked to prove the viability of the concept Overall, the concept is able to achieve the goals set, with the remaining problems that all attributes, rules and signature keys must be known to all clients, and that object keys may not be available immediately after file sharing.

References

1. Alwen, J., Coretti, S., Dodis, Y., Tselekounis, Y.: Modular design of secure group messaging protocols and the security of mls. In: Proceedings of the 2021 ACM SIGSAC Conference on Computer and Communications Security. pp. 1463–1483 (2021)
2. Barnes, R., Beurdouche, B., Robert, R., Millican, J., Omara, E., Cohn-Gordon, K.: The Messaging Layer Security (MLS) Protocol. RFC 9420 (Jul 2023). https://doi.org/10.17487/RFC9420, https://www.rfc-editor.org/info/rfc9420
3. Beurdouche, B., Rescorla, E., Omara, E., Inguva, S., Duric, A.: The Messaging Layer Security (MLS) Architecture. Internet-Draft draft-ietf-mls-architecture-11, Internet Engineering Task Force (Jul 2023), https://datatracker.ietf.org/doc/draft-ietf-mls-architecture/11/, work in Progress
4. Bhargavan, K., Beurdouche, B., Naldurg, P.: Formal Models and Verified Protocols for Group Messaging: Attacks and Proofs for IETF MLS. Ph.D. thesis, Inria Paris (2019)
5. Cohn-Gordon, K., Cremers, C., Garratt, L.: On post-compromise security. In: 2016 IEEE 29th Computer Security Foundations Symposium (CSF). pp. 164–178 (2016https://doi.org/10.1109/CSF.2016.19
6. Cohn-Gordon, K., Cremers, C., Garratt, L., Millican, J., Milner, K.: On ends-to-ends encryption: Asynchronous group messaging with strong security guarantees. In: Proceedings of the 2018 ACM SIGSAC Conference on Computer and Communications Security. p. 1802-1819. CCS

'18, Association for Computing Machinery, New York, NY, USA (2018https://doi.org/10.1145/3243734.3243747, https://doi.org/10.1145/3243734.3243747
7. Helmich, R.: Bereitstellung starker sicherheitsgarantien für cloud-dienste auf basis des messaging layer security protokolls: Konzeption und evaluation eines ende-zu-ende verschlüsselten filesharing dienstes (April 2024), written in German
8. Hu, V.C., Ferraiolo, D., Kuhn, R., Schnitzer, A., Sandlin, K., Miller, R., Scarfone, K.: Guide to Attribute Based Access Control (ABAC) Definition and Considerations. National Institute of Standards and Technology (2019).https://doi.org/10.6028/nist.sp.800-162, http://dx.doi.org/10.6028/NIST.SP.800-162
9. Kunke, J., Wiefling, S., Ullmann, M., Iacono, L.L.: Evaluation of account recovery strategies with fido2-based passwordless authentication. CoRR **abs/2105.12477** (2021), https://arxiv.org/abs/2105.12477
10. Machani, S., Philpott, R., Srinivas, S., Kemp, J., Hodges, J.: Fido uaf architectural overview. Tech. rep., FIDO Alliance (10 2020), https://fidoalliance.org/specs/fido-uaf-v1.2-ps-20201020/fido-uaf-overview-v1.2-ps-20201020.pdf
11. Park, D., Boyd, C., Moon, S.J.: Forward secrecy and its application to future mobile communications security. In: Public Key Cryptography: Third International Workshop on Practice and Theory in Public Key Cryptosystems, PKC 2000, Melbourne, Victoria, Australia, January 18-20, 2000. Proceedings 3. pp. 433–445. Springer (2000)

Anomalous Client Detection in Federated Learning

Dipanwita Thakur, Antonella Guzzo, and Giancarlo Fortino

Abstract Federated learning (FL), with the growing IoT and edge computing, is seen as a promising solution for applications that are latency- and privacy-aware. However, due to the widespread dispersion of data across many clients, it is challenging to monitor client anomalies caused by malfunctioning devices or unexpected events. The majority of FL solutions now in use concentrate on the classification problem, ignoring situations in which anomaly detection may also necessitate privacy preservation and effectiveness. The system in federated learning is unable to manage the potentially flawed behavior of its clients completely. These behaviors include sharing arbitrary parameter values and causing a delay in convergence since clients are chosen at random without knowing the malfunctioning behavior of the client. Client selection is crucial in terms of the efficiency of the federated learning framework. The challenges such as client drift and handling slow clients with low computational capability are well-studied in FL. However, the detection of anomalous clients either for security or for overall performance in the FL frameworks is hardly studied in the literature. In this paper, we propose an anomaly client detection algorithm to overcome malicious client attacks and client drift in FL frameworks. Instead of random client selection, our proposed method utilizes anomaly client detection to remove anomalous clients from the FL framework, thereby enhancing the security and efficiency of the overall system. This proposed method improves the global model convergence in almost 50% fewer communication rounds compared with widely used random client selection using the MNIST dataset.

Keywords Federated learning · Cybersecurity · Anomaly detection · Client selection

D. Thakur (✉) · A. Guzzo · G. Fortino
University of Calabria, Rende, Italy
e-mail: dipanwita.thakur@unical.it

A. Guzzo
e-mail: antonella.guzzo@unical.it

G. Fortino
e-mail: giancarlo.fortino@unical.it

N. Polatidis et al. (eds.), *Intelligent Distributed Computing XVII*, Studies in Computational Intelligence 1203, https://doi.org/10.1007/978-3-031-87639-4_8

1 Introduction

The notion of federated learning (FL) was introduced in 2016 [14]. Its core idea is to train machine learning models on independent datasets dispersed across multiple devices or parties, preserving local data privacy to some level. Since then, FL has grown rapidly and become a popular research area in the field of artificial intelligence [16]. The progress is primarily driven by three factors: the widespread use of machine learning technology, the rapid rise of big data, and global data privacy legislation. Hence, security and privacy are major characteristics of federated learning. For example, FL is susceptible to Byzantine attacks (which try to stop the model from converging) and poisoning attacks (which try to force convergence to an inaccurate model). FL is particularly vulnerable to Byzantine attacks, in which malicious users (Clients) alter trustworthy models or gradients to obstruct learning or purposely contaminate training data, causing the global model to pick up false information [20]. In a traditional FL algorithm as shown in Fig. 1, initially, the server sends global model parameters to all the randomly selected clients. The clients run their models with their dataset with the received parameters and send the updated parameter values to the server for aggregation. The aggregated parameter values are again sent by the server to the randomly selected clients. Clients vary greatly in terms of hardware

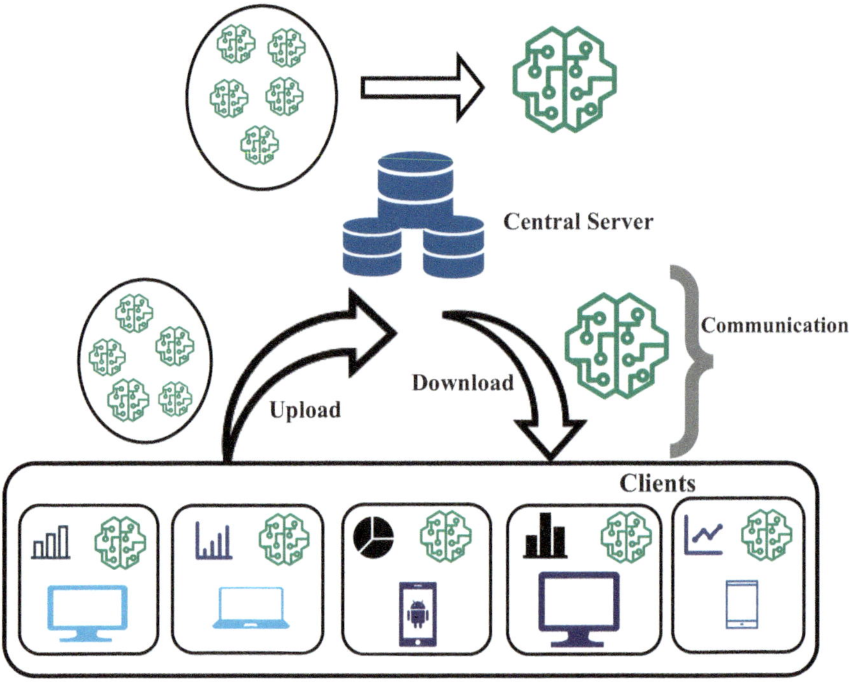

Fig. 1 Traditional FL framework

configurations and data distribution in a typical FL environment. As a result, each training round's random client sampling may not adequately take advantage of the local updates from heterogeneous clients, which could lead to decreased model accuracy, a slower rate of convergence, compromised fairness, etc. Many client selection methods have been proposed to address the FL client heterogeneity problem, with promising performance improvements. However, none of the client selection methods emphasized identifying the anomalous client and refraining them from participating in the FL network which may also lead to decreased model accuracy, a slower rate of convergence, compromised fairness, and even sabotage the whole FL network. As a solution, we can include another server for maintaining client-server binding or running any privacy-security algorithm to identify the anomalous client. Hence it will enhance the communication and computation complexity of the FL network, which researchers want to minimize for sustainable AI. An anomalous client can be one of the major privacy leakages and (or) security threat scenarios [19].

Three scenarios could result in privacy leaking and (or) security threats if a client is anomalous. Initially, the aggregator can provide the anomalous client with intermediate training updates, which they can use to examine confidential data from other client datasets [2]. Secondly, the anomalous client may submit training updates to the aggregator that are specifically tailored to probe the unique private data of other client datasets. Third, The attacker can be one of the participants in the federated learning, who adversarially modifies his parameter uploads W_i^t. Although these concerns can be mitigated by creating customized training processes, such as choosing a subset of customers for each round, FL is still exposed to a significant risk of anomalous clients leaking confidential information or any other security threat. The client can be malicious or anomalous depending on the type of attack [6]. It is pertinent to mention that whatever the type of attack it is essential to identify the malicious or anomalous clients and refrain them from taking part in the global aggregation because malicious clients can effortlessly incorporate backdoors into the combined model, all the while preserving the model's functionality for the primary goal. Moreover, the presence of anomalous clients may render model poisoning. Model poisoning, as opposed to data poisoning, entails a hostile client going for the global model directly [1, 21]. According to research by Bhagoji et al. [1], model positioning attacks have a far greater impact on the model than data poisoning attacks. The modifications made to the client's model by the malicious client, which are frequently made in FL before the new model being uploaded to the central server[4], have a direct impact on the global model, as illustrated in Fig. 2. Model poisoning affects the performance of the global model and lowers its overall accuracy by manipulating local model gradients using gradient manipulation technologies [8, 9]. For example, in Li et al. [11], when utilizing FL for image recognition, the classifier of an image model may be changed so that it assigns labels chosen by the attacker to specific areas of the image. By teaching rule modification techniques to provide attackers access to the trained model, model poisoning can also be achieved. The attacker can make their attack undetectable by changing the model's output, which allows the trained model to be updated normally [9, 10]. Wahab et al. [17] added penalty terms to reduce the discrepancies between the objective functions and appropriate weight update distribution. This enhancement led to the successful deployment of an undetected targeted model poisoning. These motivate us to propose a secure FL algorithm.

The proposed FL algorithm not only identifies the anomalous client based on the anomaly score, but this algorithm will also refrain the anomalous clients from participating in global averaging.

The main contributions are as follows:

- We propose a secure federated averaging algorithm to detect anomalous clients. Then refrain from those anomalous clients for further participation.
- The proposed algorithm is compared with the traditional FedAvg algorithm to validate its performance using iid, non-iid, and non-iid with an unequal number of features.
- To validate the proposed concept MNIST dataset of handwritten digits is used for the experiment. The proposed algorithm is secure in terms of client selection and assures convergence earlier by 50%. Hence, reduces the communication rounds.

2 Related Work

In the literature, there is substantial evidence to support the idea that federated learning frameworks enhance the security and privacy of data and networks. Furthermore, pertinent research has shown that specific characteristics of a client's private data can be revealed by models or updates provided by the client [5]. With the use of these methods, a malicious server can ascertain whether a particular data point or a sample of data points from the training distribution is used during training or not (reconstruction attacks). For instance, Melis et al. [15] discovered that during training, shared gradients can be used to collect membership data and unintentional feature leakages. Many strategies have been put forth to combat the security and privacy anomalies in FL and guard against such attacks [13, 18]. Blanchard et al. [3], examined for client changes during each loop in the context of security assaults and eliminated potentially dangerous clients when the server gathered updates. To protect against reconfiguration attacks, Gao et al. [7] suggested looking for privacy-preserving transformation functions and preprocessing training samples with them in order to guarantee the trained model performs well. In [12], the authors investigated the resilience of federated learning and suggested the PUS-FL federated learning parameter update architecture based on blockchain technology. The authors demonstrated through experiments that simulate distributed machine learning on neural networks that PUS-FL's anomaly detection algorithm performs better than traditional gradient filters, such as geometric median, Multi-Krum and trimmed mean. Furthermore, it confirms that the scalability-enhanced parameter aggregation consensus algorithm (SE-PBFT) presented in [12] reduces communication complexity to increase consensus scalability.

However, very few of them discussed anomalous client detection in a way that prevents anomalous clients from participating in aggregation and causing the model to fail to converge.

3 Proposed Algorithm

The proposed method is based on the traditional FedAvg algorithm [14]. Here we compare traditional FedAvg [14] with our proposed secure FL algorithm. The Algorithm 1 presented the traditional FedAvg algorithm.

Algorithm 1 Traditional Federated Averaging Algorithm

Require:
$C \leftarrow$ number of clients indexed by c
$E \leftarrow$ number of local epochs
$B \leftarrow$ local minibatch size
$\eta \leftarrow$ learning rate
$D_c \leftarrow$ dataset at client c
Server Execution:
Initialize w_0
for each round $t = 1, 2, \cdots$ **do**
    $m \leftarrow max(\lfloor F.C \rfloor, 1)$ ▷ F controls the global batch size
    s_t= random set of m clients
    for each client $c \in S_t$ in parallel **do**
        $w_{t+1}^{c} \leftarrow$ ClientUpdate(c, w_t)
    $m_t \leftarrow \sum_{c \in S_t} n_c$
    $w_{t+1} \leftarrow \sum_{c \in S_t} \frac{n_c}{m_t} w_{t+1}^{c}$

Client Execution:
ClientUpdate (c, w_t): ▷ Each client, c executes
$\mathcal{B} \leftarrow$ (split D_c into batches of size B)
for each local epoch i from 1 to E **do**
    for batch $b \in \mathcal{B}$ **do**
        $w \leftarrow w - \eta \nabla l(w; b)$
return w to the server

The traditional FedAvg algorithm uses a central server to coordinate training. This server houses the shared global model w_t, where t represents the communication round. Nevertheless, the real optimization is carried out locally on clients with the help of Stochastic Gradient Decent (SGD). The fraction of clients F to choose for training, the local mini-batch size B, the number of local epochs E, a learning rate η, and potentially a learning rate decay λ are the five hyperparameters of FedAvg. When using SGD for training, the parameters B, E, η, and λ are frequently utilized. But in this case, E represents the total number of iterations through the local data prior to an update to the global model. Initially, the server sends the global model, w_0

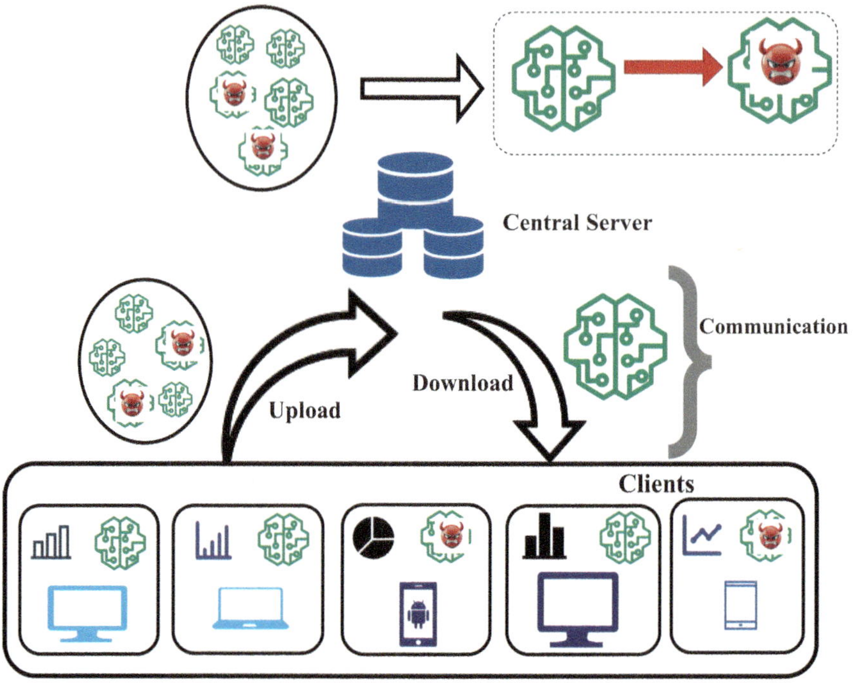

Fig. 2 Attacked FL framework with anomalous client

to all the randomly selected clients. Each communication round in FedAvg includes the following steps: The server delivers the current global model w_t to every client in S_t after choosing a subset of clients S_t, $|S_t| = F.C \geq 1$. Each client divides their local data into batches of size B and executes E epochs of SGD after updating their local models, w_t^c, to the shared model, $w_t^c \leftarrow w_t$. Lastly, clients upload their trained local models $(w_t^k + 1)$ to the server, which creates the new global model $(w_t + 1)$ by adding up all of the local models that were received. The number of local training instances determines the weighting system, as stated in Algorithm 1 on lines 12 and 13.

In the proposed secure federated averaging algorithm, we calculate the anomaly score of all the participating clients. If the anomaly score of a client is greater than the anomaly threshold value then that client is restricted to participate. Figure 2 demonstrates the FL model diagram when clients become anomalous. Even, the client is unable to send the parameter values to the global server for aggregation. Algorithm 2 represents the federated version of the proposed method step by step. We initialize the anomaly score as 1, assuming the loss as zero as clients never start their executions. The global model sends w_0 to all the randomly selected models in the first round. Then the clients update itself with the global parameters and local iterations. Each client calculates its local loss $\nabla l(w; b)$ based on their local datasets. In the further rounds, the anomaly score is calculated based on the local loss and the

Algorithm 2 Proposed Secure Federated Averaging Algorithm

Require:

$C \leftarrow$ number of clients indexed by c
$E \leftarrow$ number of local epochs
$B \leftarrow$ local minibatch size
$\eta \leftarrow$ learning rate
$D_c \leftarrow$ dataset at client c
$A^{th} \leftarrow$ threshold of anomaly score

Server Execution:

Initialize w_0, $A_0 = 1$, $S_0 = \phi$ ▷ w_0 is the global weight and A_0 is the anomaly score
if round $t = 1$ **then**
  $m \leftarrow max(\lfloor F.C \rfloor, 1)$ ▷ F controls the global batch size
  s_1= random set of m clients
for each round $t = 2, 3 \cdots$ **do**
  $\{S_t\} = \{s_1\} - \{S_a\}$
  for each client $c \in S_t$ in parallel **do**
    $w_{t+1}^c \leftarrow$ ClientUpdate(A_t, c, w_t)
    $A_{t+1}^c \leftarrow \frac{1+loss(w_{t+1}^c)}{1+\sigma_{t+1}}$
    $\sigma_{t+1} = min_c\{loss(w_{t+1}^c), c = 1, 2, \ldots, C\}$
    if $A_{t+1}^c > A^{th}$ **then**
      Add that client index in set S_a
  $m_t \leftarrow \sum_{c \in S_t} n_c$
  $w_{t+1} \leftarrow \sum_{c \in S_t} \frac{n_c}{m_t} w_{t+1}^c$

Client Execution:

ClientUpdate (c, w_t): ▷ Each client, c executes
$\mathcal{B} \leftarrow$ (split D_c into batches of size B)
for each local epoch i from 1 to E **do**
  for batch $b \in \mathcal{B}$ **do**
    $w \leftarrow w - \eta \nabla l(w; b)$
return w to the server

minimum loss of all the participating clients using the formula $A_{t+1}^c \leftarrow \frac{1+loss(w_{t+1}^c)}{1+\sigma_{t+1}}$ where $\sigma_{t+1} = min_c\{loss(w_{t+1}^c), c = 1, 2, \ldots, C\}$. Finally, after the completion of the local updates each client compares its calculated anomaly score with the threshold anomaly score. If any of the client's anomaly score is greater than the threshold then that client will be treated as an anomalous client and that client will be removed from the randomized selected client set. The threshold A_{t+1}^{th} can be chosen as, e.g., the average value or the median of the anomaly scores $\{A_{t+1}^c, c = 1, 2, \ldots, C\}$. In such a way, we can detect the anomalous client and remove it from the set of participating clients.

4 Experiment

4.1 Experimental Setup

We evaluate our proposed method with the MNIST dataset of handwritten digits, which has a training set of 60,000 examples, and a test set of 10,000 examples. It is a subset of a larger set available from NIST. The digits have been size-normalized and centered in a fixed-size image. In the FL settings, we divide the dataset in 3 different ways such as iid, and non-iid with equal number of features and non-iid with unequal number of features. It is pertinent to mention that all the clients have received the dataset as first we assign each client one shard to ensure every client has one shard of data. Next, the remaining shards are assigned randomly.

A CNN model is created using PyTorch to train with the datasets. The CNN model with two convolutional layers, each with a 5×5 kernel, 32 channels, and 2×2 max-pooling, is the image classification model taught in federated learning. It

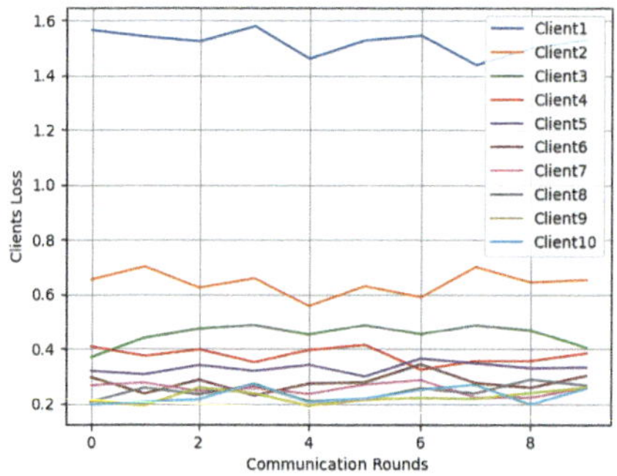

(a) Loss of each clients with iid dataset with anomalous clients

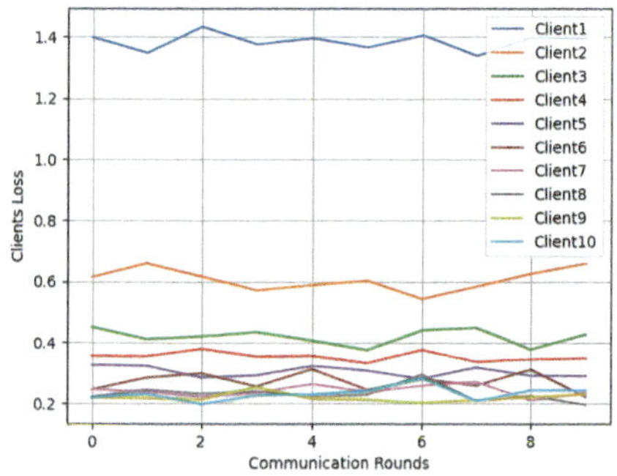

(b) Loss of each clients with non-iid dataset with equal number of features with anomalous clients

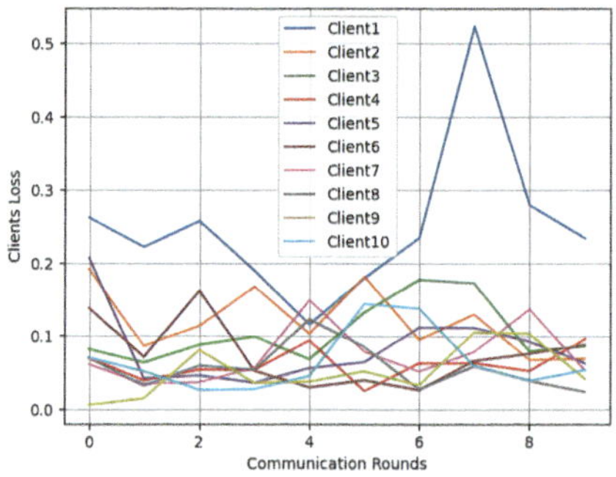

(c) Loss of each clients with non-iid dataset with unequal number of features with anomalous clients

Fig. 3 Clients loss w.r.t communication rounds with anomalous clients

is succeeded by a fully connected layer with 1024 units and a 62-dimensional softmax output layer. Within the concealed layers, ReLu activation is employed. The dataset is divided among 10 different clients. The FL was carried out for 10 communication rounds with 10 epochs in each communication round. The batch size is considered as 124 for all the experiments and the learning rate is 0.1. The parameters are fixed for all the experiments performed using the traditional federated averaging algorithm (FedAvg) and our proposed secure FL algorithm.

4.2 *Experimental Results*

The real weights of the first layer of the ML model are supplemented with a small random seed in order to generate anomalous clients. The Fig. 3 represent converging

(a) Global loss with iid dataset without anomalous clients

(b) Global loss with non-iid dataset with equal number of features without anomalous clients

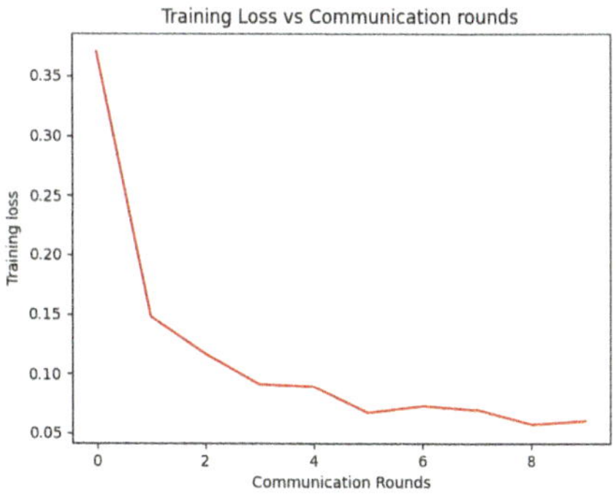

(c) Global loss with non-iid dataset with unequal number of features without anomalous clients

Fig. 4 Global Loss w.r.t communication rounds without anomalous clients

Table 1 Comparison of our proposed algorithm with the FedAvg algorithm with CNN after 10 global rounds

FL Algorithm	Data Distribution	Accuracy (%)
FedAvg	IID	95.12
	Non-IID	92.34
	Non-IID with unequal number of features	86.04
Proposed Secure FL	IID	98.54
	Non-IID	98.02
	Non-IID with unequal number of features	96.45

scenarios after manually changing the weight of the 1st layer of the two clients. From the Fig. 3, we can clearly understand that client 1 and client 2 behave abnormally. If the global model never considers those clients in the aggregation then the global model started converging after 4 communication rounds as shown in the Fig. 4 for 3 different distributions of the dataset. However, the real convergence of the global model with a non-iid dataset with an unequal number of features started after 8 communication rounds. We also compare the global accuracy with the FedAvg [14] algorithm with three different distributed datasets as shown in Table 1 after 10 global rounds. From Table 1, it is clear that our proposed algorithm is converging early with higher accuracy in comparison with traditional FedAvg. To get the 99% accuracy with traditional FedAvg, it is required to run more than 100 global iterations according to the literature. However, our algorithm converges early as the selected clients are performing well in their local iterations as we remove the anomalous clients.

5 Conclusion

In FL, data remains on edge devices for local processing. FL has several advantages over centralized machine learning (ML), including edge computing, enhanced privacy, and reduced server costs. Since the ML training process involves multiple edge devices, it's crucial to understand the state of each device. To identify anomalous clients in a distributed system, a federated secure algorithm is introduced in this paper. The main concept of this algorithm is to calculate the anomaly score of each local model based on the local loss and the minimum loss produced by all local models. Then, the algorithm uses a threshold anomaly score to identify abnormal clients, removing them from the client set for subsequent iterations. This not only detects anomalous clients but also provides a secure client selection algorithm in FL settings. The proposed method is tested on the MNIST image classification dataset, and in the future, it will be further tested with more clients using other domain-specific datasets.

Acknowledgements This work contributes to the basic research activities of the WP9.6: "AI for Green" supported by the PNRR project FAIR - Future AI Research (PE00000013), Spoke 9 - Green-aware AI, under the NRRP MUR program funded by the NextGenerationEU. **Disclosure of Interests.** The authors have no competing interests to declare that are relevant to the content of this article.

References

1. AN, B., S, C., P, M., S, C.: Model poisoning attacks in federated learning. In: In: Proc. Workshop Secur. Mach. Learn.(SecML) 32nd Conf. Neural Inf. Process. Syst.(NeurIPS). pp. 1–23 (2018)
2. Bhowmick, A., Duchi, J., Freudiger, J., Kapoor, G., Rogers, R.: Protection against reconstruction and its applications in private federated learning (2019)
3. Blanchard, P., El Mhamdi, E.M., Guerraoui, R., Stainer, J.: Machine learning with adversaries: Byzantine tolerant gradient descent. In: Guyon, I., Luxburg, U.V., Bengio, S., Wallach, H., Fergus, R., Vishwanathan, S., Garnett, R. (eds.) Advances in Neural Information Processing Systems. vol. 30. Curran Associates, Inc. (2017)
4. Cao, D., Chang, S., Lin, Z., Liu, G., Sun, D.: Understanding distributed poisoning attack in federated learning. In: 2019 IEEE 25th International Conference on Parallel and Distributed Systems (ICPADS). pp. 233–239 (2019)
5. Enthoven, D., Al-Ars, Z.: An Overview of Federated Deep Learning Privacy Attacks and Defensive Strategies, pp. 173–196. Springer International Publishing, Cham (2021)
6. Franzè, G., Tedesco, F., Famularo, D.: Resilience against replay attacks: A distributed model predictive control scheme for networked multi-agent systems. IEEE/CAA Journal of Automatica Sinica **8**(3), 628–640 (2021)
7. Gao, W., Guo, S., Zhang, T., Qiu, H., Wen, Y., Liu, Y.: Privacy-preserving collaborative learning with automatic transformation search. In: 2021 IEEE/CVF Conference on Computer Vision and Pattern Recognition (CVPR). pp. 114–123. IEEE Computer Society, Los Alamitos, CA, USA (jun 2021)
8. Hong, S., Chandrasekaran, V., Kaya, Y., Dumitraş, T., Papernot, N.: On the effectiveness of mitigating data poisoning attacks with gradient shaping (2020)
9. Jere, M.S., Farnan, T., Koushanfar, F.: A taxonomy of attacks on federated learning. IEEE Security & Privacy **19**(2), 20–28 (2021)
10. Kairouz, P., McMahan, H.B., Avent, B., Bellet, A., Bennis, M., Bhagoji, A.N., Bonawit, K., Charles, Z., Cormode, G., Cummings, R., D'Oliveira, R.G.L., Eichner, H., El Rouayheb, S., Evans, D., Gardner, J., Garrett, Z., Gascón, A., Ghazi, B., Gibbons, P.B., Gruteser, M., Harchaoui, Z., He, C., He, L., Huo, Z., Hutchinson, B., Hsu, J., Jaggi, M., Javidi, T., Joshi, G., Khodak, M., Konecný, J., Korolova, A., Koushanfar, F., Koyejo, S., Lepoint, T., Liu, Y., Mittal, P., Mohri, M., Nock, R., Özgür, A., Pagh, R., Qi, H., Ramage, D., Raskar, R., Raykova, M., Song, D., Song, W., Stich, S.U., Sun, Z., Theertha Suresh, A., Tramèr, F., Vepakomma, P., Wang, J., Xiong, L., Xu, Z., Yang, Q., Yu, F.X., Yu, H., Zhao, S.: Advances and Open Problems in Federated Learning (2021)
11. Li, M., Jin, Y., Zhu, H.: Surrogate gradient field for latent space manipulation. In: 2021 IEEE/CVF Conference on Computer Vision and Pattern Recognition (CVPR). pp. 6525–6534. IEEE Computer Society, Los Alamitos, CA, USA (jun 2021)
12. Liu, S., Shang, Y.: Federated learning with anomaly client detection and decentralized parameter aggregation. 2022 52nd Annual IEEE/IFIP International Conference on Dependable Systems and Networks Workshops (DSN-W) pp. 37–43 (2022)
13. Ma, X., Zhou, Y., Wang, L., Miao, M.: Privacy-preserving byzantine-robust federated learning. Computer Standards & Interfaces **80**, 103561 (2022)

14. McMahan, B., Moore, E., Ramage, D., Hampson, S., Arcas, B.: Communication-efficient learning of deep networks from decentralized data. In: In Proceedings of the 20th International Conference on Artificial Intelligence and Statistics; Singh, A. and Zhu, J., Eds. vol. 54, pp. 1273–1282 (2017)
15. Melis, L., Song, C., Cristofaro, E.D., Shmatikov, V.: Exploiting unintended feature leakage in collaborative learning. 2019 IEEE Symposium on Security and Privacy (SP) pp. 691–706 (2018)
16. Ng, K.L., Chen, Z., Liu, Z., Yu, H., Liu, Y., Yang, Q.: A multi-player game for studying federated learning incentive schemes. In: Bessiere, C. (ed.) Proceedings of the Twenty-Ninth International Joint Conference on Artificial Intelligence, IJCAI-20. pp. 5279–5281. International Joint Conferences on Artificial Intelligence Organization (7 2020), demos
17. Wahab, O.A., Mourad, A., Otrok, H., Taleb, T.: Federated machine learning: Survey, multi-level classification, desirable criteria and future directions in communication and networking systems. IEEE Communications Surveys & Tutorials **23**(2), 1342–1397 (2021)
18. Xie, C., Koyejo, O., Gupta, I.: Zeno: Distributed stochastic gradient descent with suspicion-based fault-tolerance. In: International Conference on Machine Learning (2018)
19. Yin, X., Zhu, Y., Hu, J.: A comprehensive survey of privacy-preserving federated learning: A taxonomy, review, and future directions. ACM Comput. Surv. **54**(6) (jul 2021). https://doi.org/10.1145/3460427, https://doi.org/10.1145/3460427
20. Zhang, C., Yang, S., Mao, L., Ning, H.: Anomaly detection and defense techniques in federated learning: a comprehensive review. Artificial Intelligence Review **57**(6), 150 (2024)
21. Zhou, Z., Xu, C., Wang, M., Ma, T., Yu, S.: Augmented dual-shuffle-based moving target defense to ensure cia-triad in federated learning. In: 2021 IEEE Global Communications Conference (GLOBECOM). pp. 01–06 (2021)

Predictive Modelling

Traditional and Bio-Inspired Feature Selection for Accurate Fraud Detection

Fidele Allarate and Franklin Tchakounte

Abstract Using credit cards is a fundamental way for financial transactions. However, their widespread use also exposes users to significant risks, particularly with the increasing prevalence of credit card fraud. To address this challenge, financial institutions and their customers must continuously adapt, as fraudsters develop increasingly sophisticated tactics. In line with this need, we have developed an effective fraud detection method emphasizing the selection of the most relevant features. Our approach leverages both bio-inspired techniques, such as genetic algorithms such as bee colony and ant colony, alongside classical methods, including sequential forward selection, Pearson's correlation, and mutual information. This research revealed that bio-inspired methods outperform classical feature selection techniques in identifying the optimal number of features for fraud detection. These selected features were subsequently used to train various ensemble learning models, including AdaBoost, Bagging, Gradient Boosting, and Stacking. Among these, the Stacking model consistently demonstrated superior performance. Notably, when the number of features ranged between 10 and 20, models based on bio-inspired methods achieved higher accuracy, F1-score, and area under the curve (AUC). However, when the feature count dropped below 10, models utilizing features selected by classical methods also delivered competitive performance, showcasing their effectiveness in scenarios with limited feature sets.

Keywords Feature selection · Bio-inspired · Fraud detection

F. Allarate (✉)
African Institute for Mathematical Sciences (AIMS), Limbe, Cameroon
e-mail: fidele.allarate@aims-cameroon.org

F. Tchakounte
Faculty of Science, University of Ngaoundere, Ngaoundere, Cameroon

N. Polatidis et al. (eds.), *Intelligent Distributed Computing XVII*, Studies in Computational Intelligence 1203, https://doi.org/10.1007/978-3-031-87639-4_9

1 Introduction

Credit card transactions have become fully integrated into our daily lives. We can use it for online shopping, paying bills, or dining out, but credit cards offer unparalleled convenience and flexibility [19]. Their use eliminates the need to carry cash, making them practical financial tools that make our lives easier. However, this widespread use of credit cards also places individuals and businesses at significant risk, with the increasing emergence of credit card fraud. It is a challenge and represents a major concern for financial firms and their customers as it takes on a global scale and affects millions of people each year. In 2023, the Federal Trade Commission (FTC) estimated that credit card fraud resulted in losses of 1.9 billion USD in the United States. In addition, fraudsters are continually evolving their tactics, using sophisticated techniques to compromise credit card data. And this advance of fraudsters is noted by Matt [32], who estimates that credit card fraud will reach 12.5 billion USD by 2025 in the United States and reach 43 billion USD by 2026 globally. Therefore, it is urgent for researchers to find effective solutions to reduce these alarming figures. To combat credit card fraud, many researchers have developed techniques using machine learning (ML). These ML methods leverage data to develop, test, evaluate, and refine fraud detection strategies. However, due to privacy, security, legal, and policy constraints, access to detailed information on fraud patterns is often difficult or impossible for external analysts. Feature selection can help overcome these issues by allowing models to operate efficiently with fewer training data.

Several feature selection methods have been proposed, grouped into conventional/traditional and biology-inspired approaches [1, 2, 4, 11, 13, 13, 18, 20–22, 24, 27, 28, 30]. The first group aims to find a subset of features that provides the best model for a given dataset. The second group is inspired by the biological evolution of nature to select the optimal number of features. To select the optimal technique, it is necessary to examine feature selection methods in terms of the constitution of subsets of features, which provides accurate fraud detection.

In line with that, this work seeks to explore and evaluate various feature selection methods and their extracted subsets of features, including the traditional and bio-inspired ones, to identify the best approach to achieve accurate fraud detection.

The rest of the paper is organized as follows. In Sect. 2, related works are discussed. In Sect. 3, the methodology is described. In Sect. 4, experiments are presented and results are discussed. The document is concluded in the last section, by providing future works.

2 Related Works

This section presents a review of the literature on fraud detection employing feature selection and bio-inspired algorithms. Numerous studies have explored diverse methodologies in this area, showcasing significant advancements and highlighting various techniques' efficacy. An empirical study by [13] proposed a novel methodology for credit fraud detection (CFD) based on machine learning. The authors utilized an Informative Variable Identifier (IVI) to differentiate informative, redundant, and noisy variables. Subsequently, a set of recurrent filters was applied, followed by classification using advanced machine learning techniques such as gradient boosting, support vector machines (SVM), linear discriminant analysis (LDA), and linear regression. Their approach achieved a 76% accuracy rate, representing an improvement of over 4% compared to previous works on real-world datasets. In another extensive study by [24], the authors investigated the impact of feature selection on the accuracy of various machine learning models for fraud detection. Using a correlation matrix for feature selection, they implemented random forest (RF) and logistic regression (LR) models. Their findings revealed that the random forest classifier with feature selection outperformed others, achieving a 98% accuracy rate, demonstrating the enhancement feature selection brings to model performance. A comparative study by [20] examined the differences between quantum and classical feature selection techniques and their impact on Quantum Machine Learning (QML) classifiers. They found that quantum algorithms identified optimal feature subsets distinct from those selected by classical methods, underscoring the complementary role of quantum support vector machines (QSVM) in feature space exploration. Several studies [1, 2, 13, 18, 21, 22, 30] have relied on hybrid feature selection techniques to improve credit card fraud detection. Others, such as [16, 26], employed meta-heuristic optimization approaches to achieve highly accurate detection. For instance, [27] introduced a Binary Emperor Penguin Optimization (BEPO) algorithm integrated with an Optimal Gated Recurrent Unit (OGRU) model, achieving accuracies of 94.78% and 94.16% on the German Credit and Credit Card Fraud Detection datasets, respectively. In addition, feature selection techniques such as Filters, Wrappers, and Embedding were explored in [4]. Among the classifiers evaluated, Support Vector Data Description (SVDD) achieved the highest classification metrics, with 93% accuracy, 90% precision, 97% sensitivity, and an F-measure of 93%. Interestingly, classifiers such as Logistic Regression (LR), Decision Tree (DT), SVM, and k-Nearest Neighbors (k-NN) were unaffected by feature selection methods, maintaining the same performance as before. Only the SVDD and RF models constructed using selected feature subsets outperformed those based on the original dataset. Finally, a recent study by [11] employed the Grasshopper Optimization Algorithm to enhance feature selection for fraud detection. Their results highlighted the multilayer perceptron (MLP) as a top performer, achieving precision (0.942), recall (0.891), F-measure (0.915), and an accuracy rate of 95%. These findings demonstrate the promise of bio-inspired algorithms and advanced feature selection methods in improving fraud detection systems. In summary, feature selection techniques are revealed to be helpful to improve performance. Also, the selection of optimal number of features is undoubtedly required to enhance accuracy. Table 1 summarizes the related works. The first column refers to works. The second column refers to feature selection algorithms used in the study. The third column refers to key performance results. The fourth column represents the condition that guides the choice of the feature selection algorithm. When it is NO, it means that the paper has not determined such condition. The last column refers to the dataset that has been used in the paper.

Table 1 Related works

Authors	Aproaches	Contributions	Conditions	Year
Anurag Jain et al. [2]	Firefly Algorithm	Increase the accuracy	NO	2019
All those work [9, 17, 35]	fruit fly algorithm (FFA)	Increase the accuracy	NO	2020
Anurag Jain, Ajeet Singh [1]	Filter and Wrapper	Improve the accuracy	NO	2021
Victoria Priscilla, Padma Prabha [31]	Mutual Information and Recursive Feature Elimination	Reduce the accuracy	NO	2021
Kannan et al. [29]	Information Gain	Improve the accuracy	NO	2021
Santiago Moral-Rubio et al. [13]	Informative Variable Identifier (IVI)	Increase the accuracy	NO	2022
P. Rajesh Pandurangan [27]	Metaheuristic method	Increase the accuracy	NO	2022
Ibomoiye Domor Mienye and Yanxia Sun [21]	Information Gain(IG) and Genetic Algorithm (GA)	improve the sensitivity and specificity	NO	2023
Prabhakaran et al. [30]	Oppositional Cat Swarm Optimization- based(OCSO)	Increase the accuracy	NO	2023
Boluwatife [24]	Correlation matrix	Increase the accuracy	NO	2024
Parveen Gupta [11]	Grasshopper algorithm	Increase the accuarcy	NO	2024

We can clearly see that none of these works identified a condition to guide which category of feature selection to choose. Complementary to these works, this study proposes first to investigate which of traditional and bio-inspired is better for credit card detection, second to explore the subset of features that provides interesting results in both categories of feature selection, and third to find a minimum number of features which helps to choose the category of feature selection.

3 Methodology

Features are key piece in ML modeling. Basically, model predictions directly depend on the quality of features. As a result, in ML, a special emphasis is placed on feature engineering and feature selection. To know which features contribute more to the model and, provided that, select or create new features, we would need to measure their importance [23]. In this study, the following steps summarized in Fig. 1 are performed to select the relevant features for accurate fraud detection:

1. Constitution of credit card transaction datasets with fraud labels;
2. Application of feature selection algorithms to datasets;
3. Train ML classifiers (Ada Boost, Bagging, Gradient Boosting and Stacking) on selected features;
4. Evaluate models the models;
5. Compare algorithms based on accuracy, F1, and AUC-ROC.

In the following sections, every step is detailed.

3.1 Acquisition of Data

The first phase of our work involves actively searching for a recent dataset containing information about credit card transactions. This step is crucial because it allows us to take into account new emerging fraud techniques. We place great importance on the relevance and reliability of data to ensure the accuracy of our analysis. This part constitutes the essential basis on which our comparative study of fraud detection methods is based. Our data set must also have a variable that represents fraud and normal transactions. Then we will undertake the data cleaning process to ensure its quality and reliability. If any data has anomalies or inconsistencies, we will identify and correct them through data cleaning. In addition, we will assess the issue of class imbalance, where some categories may be underrepresented compared to others in our dataset. In the case of data imbalance, we will resolve this problem, using techniques such as oversampling to ensure more balanced and accurate predictions. This reassured us to have balanced data set for the rest of our work.

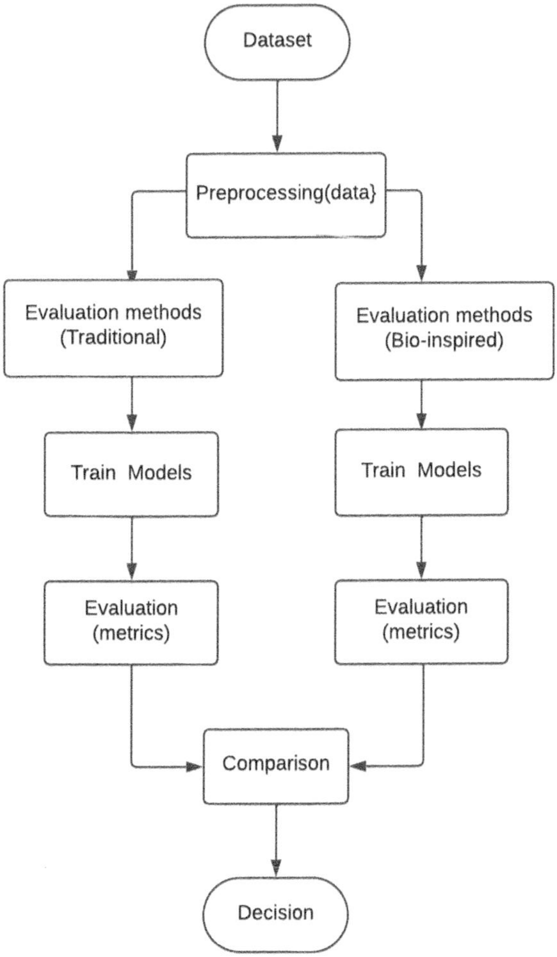

Fig. 1 Methodology

3.2 *Feature Selection*

Features are a central and fundamental element of the field of machine learning. Indeed, the quality of the features directly influences the model predictions. Thus, feature engineering and selection are crucial aspects of machine learning. To determine which features contribute the most to the model and decide whether to select or create new features, it is necessary to measure their importance in one way or another [23].There are several ways of processing features, we can select the features, transform the features in order to have new features based of course on our current features. We decided to consider in our work the selection of features, without transformations of the features, to keep the structure and the relationship between

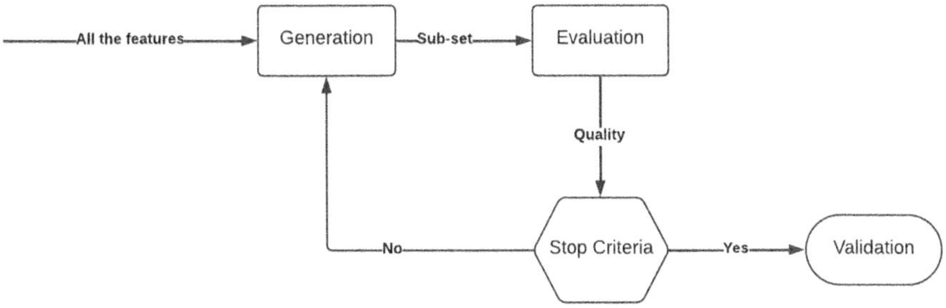

Fig. 2 The feature selection process

the variables; this allows them to maintain their purity in the detection of fraud with real variables.

Formulation of problem: We consider that $F = \{f_1, f_2, \ldots, f_n\}$ is our set which contains the features; the number of ours features is n.
Let also E_v be a function, which can evaluate a subset of features. We are looking for the subset that will give us the greatest value E_v. We can use this objective function according to the best subset S ($S \subseteq F$) [7]:

$$E_v(S) = \max_{Z \subseteq F} E_v(Z), \tag{1}$$

where $|Z| = n$ and n is a number predefined of the subset generation methods. A general procedure for a feature selection method is illustrated in Fig. 2.

Following this outline, we start with a broad set of potential characteristics. Then, we evaluate these characteristics according to their relevance and importance, using techniques for feature selection. Finally, we will decide to select an optimal number of features that increase the performance of the models while reducing complexity and computational time. There are two kind of feature selection: Traditional and Bio-inspired.

Traditional Algorithms: Traditional Algorithms are divided into three main parts [6, 14]: wrappers, filters and embedding. Our approach is based on some of the most popular methods.

1. **Wrappers:**
   When using the wrapper-based approach for feature selection, we must use a classification model. We need to select a subset of features from the set of all the features by evaluating the model and considering the features that maximize accuracy. Among many alternatives, sequential feature selectors constitute an important family of search algorithms. Basically, they add or remove one feature at a time and count the performance of a model until a subset of features of a desired size S (S < I) is achieved [15]. There are two variations: Sequential Forward Selection (SFS) and sequential backward selection (SBS). Since the ideas of SFS and SBS are quite similar, we will only disassemble SFS.

Given a set of features F:

$$F = \{f_1, f_2, \ldots, f_I\} \tag{2}$$

where $f_1 \ldots f_I$ are the features belong to the set F. The aim is to select a subset of features which maximize the accuracy.

$$\Phi_S = \{\phi_1, \ldots, \phi_s, \ldots, \phi_S\} \tag{3}$$

with $\phi_s \in F$ and $S = 0, 1, 2, \ldots, I$, then $S < I$.
Initially, for $S = 0$, $\Phi_0 = \emptyset$.

$$R = F \backslash \Phi_S \tag{4}$$

Let $Q(\cdot)$ stand for the quality measure of a classifier (accuracy in our work). The feature that is associated with the best classifier performance can be established as follows [15]:

$$\hat{\phi} = \arg \max_{r=1,\ldots,|R|} Q(\Phi_S \cup \phi_r), \tag{5}$$

$$\Phi_{S+1} = \Phi_S \cup \hat{\phi} \tag{6}$$

2. **Filters:**
   This method uses a feature ranking function to choose the best features. The ranking function gives a relevance score based on a sequence of examples. The most relevant feature has a high rank. We must therefore choose a fixed number of features having the highest rank, or according to a variable number of features above a predefined threshold value of the rank. Talk to us to discuss some of the most popular ones.

   (a) **Mutual Information:**
   The mutual information (MI) measures how much information of a random variable is contained by another variable. It is defined when the variables are continue by:

$$I(X, Y) = \int\int p(x, y) \times log \frac{p(x, y)}{p(x) \times p(y)} dx\, dy \tag{7}$$

   where X and Y are the variables; if the variables are categorical, we must use this formula:

$$I(X, Y) = \sum\sum p(x, y) \times log \frac{p(x, y)}{p(x) \times p(y)} \tag{8}$$

For each variables we should calculate this value according to our target variables. After calculating this information, the variables will be arranged in descending order and the k best will be selected.

(b) **Pearson Correlation:**
It is a statistic measure used to establish a linear correlation between two variables. For the i-th feature, i = 1, …, I, of L input records and the available data targets t_l, the correlation between the i-th input and the output can be defined as follows [15].

$$r_i = \frac{\sum_{l=1}^{L}(x_{li} - \bar{x}_i)(t_l - \bar{t})}{\sqrt{\sum_{l=1}^{L}(x_{li} - \bar{x}_i)^2 \sum_{l=1}^{L}(t_l - \bar{t})^2}} \tag{9}$$

where $\bar{x}_i = \frac{1}{L}\sum_{l=1}^{L} x_{li}$ is the mean over the i-th feature, while $t = \frac{1}{L}\sum_{l=1}^{L} t_l$ denotes the mean over the targets of all x_l records for l = 1, …, L. If x_l and t_l are linearly dependent, $r_i = \pm 1$; $r_i = 0$, if the variables are not correlated. On the basis of decreasing values of $|r_i|$, the ranking of feature importance can be established [15].

Bio-Inspired Algorithms: Metaheuristic optimization methods are obviously inspired by nature. Their mechanism and abilities are extraordinarily magical and mysterious and researchers have focused on mapping the natural phenomenon onto intelligence algorithms [3]. Their capabilities have been described to solve the complex problem independently of elementary populations and initial parameters with little or no knowledge of the feature space. Bio-inspired algorithms are one of the most popular categories of these meta-heuristics. In our work we decided to use Genetic Algorithm, Ant colony and Bee colony.

1. **Genetic Algorithm**
We generated a population; each individual in this population has exactly all the features. These features are encoded by 0 or 1, where 0 indicates that the feature which is at this position is not selected and 1 indicates that this feature has been selected. The objective is to determine the most relevant set of p descriptors for predicting a physicochemical property, the fitness function should estimate the accuracy of a prediction method when only the p feature are used. The fitness function is the mean square error of prediction (MSE) [3].

$$F(P_{z_{1,k}}, Z_{2,k}) = \frac{1}{n_{2(\hat{x}_i, y_i) \in Z_{2,k}}} \sum (y_i - P_{z_{1,k}}(\bar{x}_i))^2 \tag{10}$$

Where:

- Z is a matrix that represents a compound dataset;
- PZ is a predictor method trained with the dataset Z [3];
- Z_1 and Z_2 are compound databases;
- $Z_{i,k}$ is a filtered dataset according to the descriptor selection encoded in the kth individual;
- $\bar{x}_i$ is a vector that represents the values of the descriptors for the ith compound of a given dataset.
- y_i is the target value for the ith compound of Z.

After fitness evaluation, new population is created using Crossover and/or Mutation. Until to find a stop criteria which is number of iterations in our work.

2. **Ant Colony** seek to minimize the redundancy of features by using the notion of similarity within features. We formulate the problem as a graph $G = (F, E)$ where $F = \{F_1, F_2, \ldots, F_n\}$ is the set of features and $E = \{(F_i, F_j)\}$ indicates the edges of the graph. We used correlation to represent the connection of the boundary between two features. The similarity between to feature F_i and F_j is found by using [33]:

$$sim(F_i, F_j) = \left| \frac{\sum_1^p = 1(a_i, b_j)}{(\sqrt{\sum_1^p = 1^{a_1^2}})(\sqrt{\sum_1^p = 1^{b_1^2}})} \right| \tag{11}$$

Where $F_i = \{a_1, a_2, \ldots, a_p\}$, $F_j = \{b_1, b_2, \ldots, b_p\}$ are the vectors and p is the number of others features. This similary has two values: 0 means that the features are not similar and 1 does mean, those features similar.

In each node, there is initial quantity of pheromone allocated, and the ants are randomized in the nodes. Those ants are moving in each iteration until to find the stop criteria. The rule of transition is concerning the selection of feature with high values of pheromone and small similarity with the feature in this node. At end of iteration we update the value of pheromone. Global updating is done using:

$$\tau_{ij}(t + n) = \rho\, \tau_{ij}(t) + \Delta\tau_{ij} \tag{12}$$

And the $\Delta\tau_{ij}^k$ is the quantity of pheromone:

$$\Delta\tau_{ij} = \sum_{k=1}^{m} \Delta\tau_{ij}^k \tag{13}$$

The process is repeated until a certain number of iterations is achieved. The features are grouped by their pheromone values in lower order. The top selected features with highest pheromone values are selected as the final feature subset. But concretely the new feature j is chosen using the following equation [33]:

$$j = \arg\max_{u \in j_i^k} \left\{ (\tau_{ij}^{\alpha})(\eta(F_i, F_u))^{\beta} \right\} \quad if \ \ q \leq q_0 \tag{14}$$

where j_i^k is the unvisited feature set, τ_u is the pheromone assigned to the feature u; the parameters α and β are used to determine the relative importance of the intensity of the trace and the visibility in the construction of a solution. the inverse of similarity is represented by [33]:

$$\eta(F_i, F_j) = \frac{1}{sim}(F_i, F_u) \tag{15}$$

The Mutual Information ranks the features according to the values of information.

3. **Bee Colony Optimization** is also possible to solve this problem by using Bee Colony Optimization (BCO). To do this, we will start with [10]:

   (a) Initialization:We consider each food source as a probable solution which is described as a vector:

   $$\mathbf{X} = (x_{i,1}, x_{i,2}, \ldots, x_{i,D}) \tag{16}$$

   We can directly generate by:

   $$x_{i,j} = x_j^{min} + rand(0, 1)(x_j^{max} - x_j^{min}) \tag{17}$$

   where $i = 1, 2, \ldots, N$ is the number of real food sources and $j = 1, 2, \ldots, D$ is the dimensionality of the search space. $x_{i,j}$ represents the jth dimension of x_i. x_{max} and x_{min} are the maximum and minimum boundary values, respectively.

   (b) Employed bee phase: Those ants are in each food source and they must change their position if there is one new position better than their current position according to their neighbors. The new food source is produced by [10]:

   $$x'_{i,j} = x_{i,j} + \phi_{i,j}(x_{i,j} - x_{k,j}) \tag{18}$$

   If $f(x'_i) > f(x_i)$, then x'_i replaces x_i.

   (c) Onlooker bee phase: These bees obtain information regarding the amount of nectar (fitness value) and the position of food sources from the employed bees. They select a food source close to them using this formula:

   $$P_i = \frac{fitness_i}{\sum_{j=1}^{SN} finess_j} \tag{19}$$

   (d) Scout bee phase: Each source has a counter which is zero at the beginning. If the counter holding the number of trials exceeds the predefined threshold value, its corresponding food source will be abandoned and replaced by a new food source.

After reviewing the algorithms for feature selection, we will deepen our analysis by exploring the different models that we employ in our work. This exploration will allow us to understand how each model processes data and contributes to accurate fraud detection.

3.3 Training

We used ensemble learning algorithms. Ensemble learning is one of the categories of machine learning models, generally used for global decision-making. It is adopted in this work because they use several learning algorithms to obtain better performance than those of the algorithms individually [25]. We train ML classifiers (Ada Boost, Bagging, Gradient Boosting, and Stacking) on selected features. After selecting the relevant features, we used them to validate the machine models. These models are then trained to detect fraud cases effectively. This process helps optimize our system's ability to spot suspicious behavior and protect users from fraudulent activity. In addition, we will also learn to recognize fraud patterns accurately, which strengthens the security of our platform (approach).

Ada Boost: AdaBoost is an ensemble learning method using the notion of boosting. It can normally be used well to solve the binary classification problem [8]. Consider a set $S = (x_1, y_1), (x_2, y_2), \ldots, (x_n, y_n)$ where $\forall i,\ y_i \in \{0, 1\}$ and a sequence of hypotheses group together into a set H, which we will choose T hypotheses H. H then makes a decision using the individual hypotheses $h_1, \ldots, h_T$ by this formula [8]:

$$H(x) = \sum_{i=1}^{T} \alpha_i h_i(x) \qquad 1.$$

In fact, this method involves the formation of a linear combination of the different hypotheses available, where each hypothesis contributes to a specific part of the solution. The algorithm then selects these hypotheses sequentially, gradually evaluating and adjusting them to achieve the desired result. Let H_t be the classifier formed by the first t hypotheses and:

$$H_t(x) = \sum_{i=1}^{t} \alpha_i h_i(x) = H_{t-1}(x) + \alpha_t h_t(x), \tag{20}$$

where $H_0(x) = 0$, it is the empty ensemble and will have always output 0 [8].

Bagging: Consider that we have (X_i, Y_i) with $i = 1, \ldots, n$, where X_i denotes the d-dimensional of the variables and the target variable is $Y_i = \{0, 1\}$ for the binary classification. The target function of interest is the multivariate function $\mathbb{P}[Y = j | X = x]$ $(j = 0, \ldots, J - 1)$ in the case of classification, and the function estimator is [5]:

$$\hat{g} = h_n((X_1, Y_1), \ldots, (X_n, Y_n)), \tag{21}$$

where the function $h_n(\cdot)$ is the estimation function. The Bagging method is defined using these steps [5]:
Step 1: Define by randomly drawing a bootstrap sample $(X_1^*, Y_1^*), \ldots, (X_n^*, Y_n^*)$ with replacement from the data $(X_1, Y_1), \ldots, (X_n, Y_n)$.
Step 2: Compute the estimator $\hat{g}^*$ by the plug-in principle:

$$\hat{g}^* = h_n((X_1^*, Y_1^*), \ldots, (X_n^*, Y_n^*)).$$

Step 3: Repeat steps 1 and 2 M times, with M taken between 50 or 100.

Gradient Boosting: Gradient boosting is based on boosting with the aim of creating strong ensembles. It works primarily using decision trees as a basis for producing a robust ensemble classifier; this is why we give it the name gradient-boosted decision tree (GBDT) [34]. Let S a set such that:

$$S = \{(X_i, Y_i)\}. \tag{22}$$

The goal of the gradient boosting technique is to find an approximation $\hat{F}(X)$, of a function $F^*(X)$ which is naturally considered as a mapping of the variables X to their response variables Y. The GBDT creates another approximation of $\hat{F}(X)$ as [34]:

$$F_m(X) = F_{m-1}(X) + \rho_m h_m(X), \tag{23}$$

where ρ_m is the weight of the m_{th} function $(h_m(X))$. The constant approximation of $F^*(X)$ is [34]:

$$F_0(X) = \text{argmin}_\alpha \sum_{i=1}^{N} L(Y_i, \alpha). \tag{24}$$

Stacking: Stacking is an ensemble learning technique that differs from the others in its way of combining machine learning algorithms. Unlike other ensemble methods such as majority voting or prediction averaging, stacking trains an additional machine learning algorithm, often called a *meta-model* or *meta-classifier*, to combine the predictions of different members of the ensemble.

Figure 3 illustrates the structure of the stacking model.

This requires building models based on several classic machine learning algorithms called level 0 models and a meta-learning algorithm that trains another model to combine predictions from the base models. This meta-model is called the level 1 model. The idea is that the basic level 0 algorithms are trained using the training dataset and given out-of-sample information or unseen data [34];

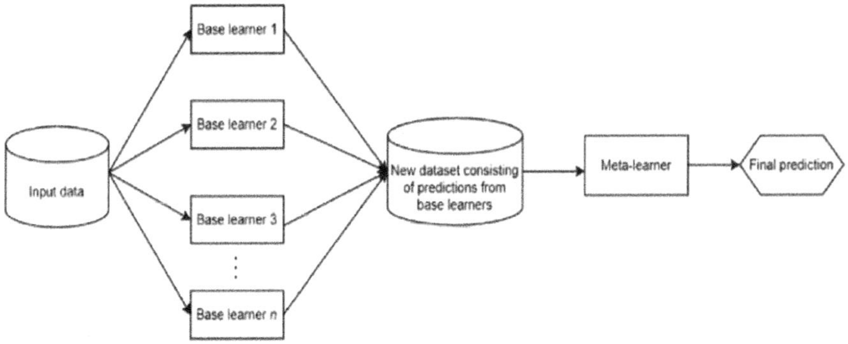

Fig. 3 Stacking model [34]

3.4 Evaluation

Training models are evaluated based on four (04) metrics [12].

1. **Accuracy**: Accuracy is a fundamental metric used to evaluate the performance of a model, especially in classification tasks. This metric measures the percentage of correct predictions made by the model among all predictions. More precisely, it evaluates how often the model's predicted class label matches the actual class label. The formula for calculating accuracy is:

$$Accuracy = \left(\frac{\text{total of correct predictions}}{\text{total of predictions}}\right) \times 100. \tag{25}$$

2. **Recall**: Recall is a metric that measures how often a machine learning model correctly identifies positive instances (true positives) from all the actual positive samples in the dataset. Recall is often used in contexts where the goal is not to miss true positive examples. For example, in applications such as disease detection (such as cancer) or fraud detection, it is crucial to detect as many positive cases as possible, even if this means accepting a certain number of false positives.

$$Recall = \frac{\text{true positive}}{\text{true positive} + \text{false negative}}. \tag{26}$$

3. **F-1 Score**: F1-score is a critical metric in machine learning, especially for classification tasks where data may be imbalanced or when both precision and recall are important. It provides a balanced measure of a model's accuracy by combining two aspects (Precision and Recall). F1-score is calculated using this following formula:

$$F1 - score = 2 \times \frac{\text{precision} \times \text{recall}}{\text{precision} + \text{recall}}. \tag{27}$$

And the precision is given as follows:

$$Precision = \frac{\text{true positive}}{\text{true positive} + \text{false positive}}. \tag{28}$$

4. **AUC**: Area Under Curve (AUC) is a metric used to evaluate the robustness of the model based on its ability to distinguish between positive and negative samples. AUC is graphically represented by the Receiver Operating Characteristic (ROC), which is the area of the zone delimited by the curve [12].

3.5 Comparison

In this step, the different feature selection techniques are compared based on the performance results obtained after training the chosen ensemble learning algorithms. The performance of the models is performed using the previously defined metric, which allows to quantify their effectiveness. This comparison allows to identify the most relevant selection method, based on the impact it has on the accuracy of the model. Thus, the objective is to determine the most reliable technique.

4 Experiments and Results

We rigorously applied the methods mentioned in the methodology. Each step of our approach was followed methodically, from data analysis to evaluation of results, ensuring precise implementation in line with best practices. After obtaining the results, we carried out a detailed comparison of the performance of each one, using relevant metrics to assess their effectiveness. This comparison allowed us to make an informed final decision, based on objective results and rigorous analysis.

4.1 Data Collection

The dataset[1] taken from Kaggle has been used in our study to detect credit card fraud. It contains credit card transactions made by European cardholders in the year 2023 with sensitive information removed to ensure privacy and compliance with ethical guidelines. It also comprises over 550,000 records, and the data has been anonymized to protect cardholders' identities. This dataset has been created with the goal to facilitate the development of fraud detection algorithms and models that identify potential fraudulent transactions. It contains 31 features such as

[1] https://www.kaggle.com/datasets/nelgiriyewithana/credit-card-fraud-detection-dataset-2023.

id, V_1, V_2, V_3, V_4, V_5, V_6, V_7, V_8, V_9, V_{10}, V_{11}, V_{12}, V_{13}, V_{14}, V_{15}, V_{16}, V_{17}, V_{18}, V_{19}, V_{20}, V_{21}, V_{22}, V_{23}, V_{24}, V_{25}, V_{26}, V_{27}, V_{28}, *amount* and *Class* that is the boolean target variable. All these variables are numerical; We checked if there are any missing values, but we did not have missing values. This data is suitable for research in the field of fraud detection, it is particularly processed and we do not have much to do, so as not to deviate from the objective of the dataset. The dataset is balanced and clean. The current concern is applying feature selection algorithms. We used only 29 features apart from *id* and *amount*. We conducted an in-depth statistical study of our dataset to gain a clear understanding of the data before proceeding with our study. The descriptive values in Table 4 highlight significant variations between different categories.

We remark that the mean of all variables, ranging from V_1 to V_{28}, is approximately zero. This close proximity to zero suggests that the data has been effectively standardized. Successful standardization is crucial in data analysis, as it ensures that our variables(features) are on a comparable scale. This process helps to eliminate biases that could arise from different ranges of variables. Therefore, the results indicate that the preprocessing steps were executed correctly, contributing to more reliable analytical outcomes. Furthermore, the standard deviation is uniformly equal to 1 for all variables, reinforcing the idea of effective standardization of this dataset. A standard deviation of 1 suggests that the spread of the data is consistent, which is important for maintaining the integrity of the analysis. Although these variables are well standardized, it is essential for us to select the most relevant features in order to detect fraud with increased accuracy; while the standardization of our dataset lays a solid foundation, the next critical step involves meticulously choosing the features that will drive our fraud detection efforts forward.

4.2 Extraction of Feature Subsets

The core focus of our work revolves around rigorously evaluating various feature selection algorithms. It is crucial to understand that not all algorithms operate in the same manner, leading to the likelihood that the resulting subsets of selected features will vary significantly between them. To address this, we will conduct a series of experiments with these algorithms to analyze their performance and characteristics. By doing so, we aim to gain a comprehensive understanding of how each algorithm functions, which will enable us to identify the nuances of the feature subsets they produce.

We experimented one by one with the feature selection algorithms that are in our approach; for each of them, we applied it to the dataset to select the features and train, then evaluate our models.

We started by applying Sequential Feature Selection (SFS) on this dataset. This algorithm works with a machine learning model (classifier), we used logistic regression; we also investigated this algorithm with different accuracy values (0.9, 0.8, 0.7); but our goal is to have the features that increase the accuracy of models. The

	count	mean	std	min	25%	50%	75%	max
id	568630.0	2.843145e+05	164149.486122	0.000000	142157.250000	284314.500000	426471.750000	568629.000000
V1	568630.0	-5.638058e-17	1.000001	-3.495584	-0.565286	-0.093638	0.832658	2.229046
V2	568630.0	-1.319545e-16	1.000001	-49.966572	-0.486678	-0.135894	0.343555	4.361865
V3	568630.0	-3.518788e-17	1.000001	-3.183760	-0.649299	0.000353	0.628538	14.125834
V4	568630.0	-2.879008e-17	1.000001	-4.951222	-0.656020	-0.073762	0.707005	3.201536
V5	568630.0	7.997245e-18	1.000001	-9.952786	-0.293496	0.081088	0.439737	42.716891
V6	568630.0	-3.958636e-17	1.000001	-21.111108	-0.445871	0.078718	0.497788	26.168402
V7	568630.0	-3.198898e-17	1.000001	-4.351839	-0.283533	0.233366	0.525955	217.873038
V8	568630.0	2.109273e-17	1.000001	-10.756342	-0.192257	-0.114524	0.047299	5.958040
V9	568630.0	3.998623e-17	1.000001	-3.751919	-0.568745	0.092526	0.559262	20.270062
V10	568630.0	1.991314e-16	1.000001	-3.163276	-0.590101	0.262614	0.592460	31.722709
V11	568630.0	-1.183592e-16	1.000001	-5.954723	-0.701449	-0.041050	0.747773	2.513573
V12	568630.0	-5.758017e-17	1.000001	-2.020399	-0.831133	0.162052	0.744672	17.913556
V13	568630.0	-5.698037e-18	1.000001	-5.955227	-0.696667	0.017608	0.685605	7.187486
V14	568630.0	-4.078595e-17	1.000001	-2.107417	-0.873206	0.230501	0.751822	19.169544
V15	568630.0	2.649087e-17	1.000001	-3.861813	-0.621249	-0.039256	0.665406	14.532202
V16	568630.0	-1.719408e-17	1.000001	-2.214513	-0.716265	0.134026	0.655606	46.652906
V17	568630.0	-3.398829e-17	1.000001	-2.484938	-0.619491	0.271641	0.518224	6.994124
V18	568630.0	-5.837989e-17	1.000001	-2.421949	-0.556046	0.087294	0.544389	6.783716
V19	568630.0	2.479146e-17	1.000001	-7.804988	-0.565308	-0.025979	0.560116	3.831672
V20	568630.0	-1.579456e-17	1.000001	-78.147839	-0.350240	-0.123378	0.248216	29.872812
V21	568630.0	4.758361e-17	1.000001	-19.382523	-0.166441	-0.037431	0.147979	8.087080
V22	568630.0	3.948640e-18	1.000001	-7.734798	-0.490489	-0.027329	0.463882	12.632511
V23	568630.0	6.194741e-18	1.000001	-30.295450	-0.237629	-0.059689	0.155715	31.707627
V24	568630.0	-2.799036e-18	1.000001	-4.067968	-0.651580	0.015901	0.700737	12.965639
V25	568630.0	-3.178905e-17	1.000001	-13.612633	-0.554148	-0.008193	0.550015	14.621509
V26	568630.0	-7.497417e-18	1.000001	-8.226969	-0.631895	-0.011892	0.672888	5.623285
V27	568630.0	-3.598760e-17	1.000001	-10.498633	-0.304961	-0.172911	0.334023	113.231093
V28	568630.0	2.609101e-17	1.000001	-39.035243	-0.231878	-0.013930	0.409590	77.255937
Amount	568630.0	1.204196e+04	6919.644449	50.010000	6054.892500	12030.150000	18036.330000	24039.930000
Class	568630.0	5.000000e-01	0.500000	0.000000	0.000000	0.500000	1.000000	1.000000

Fig. 4 Data description

best accuracy value in our work is 0.9; and this gave us 21 features, but the other accuracy values gave us all the features in the dataset, so this subset of 21 features is the best.

Secondly, we also applied Recursive Feature Elimination (RFE) with the same classifier as for SFS (Logistic Regression), but with the difference we must choose, set the number of features (K) that we want to have as output from this algorithm; to properly investigate our algorithm, we set our K to 5, 10, 15.

We continued with Mutual Information (MU) by using a function (*mutual_info_classif*), already existing in the sklearn library. The mutual information is calculated between each characteristic and the target variable (Class), then the features are classified according to their mutual importance with the target variable. Features with the highest mutual information values are often considered the most informative for predicting the "Class" and are therefore selected for model building. This algorithm will necessarily give us all the features ranked in their , it is up to us to select the best ones in our work.

The final algorithm as traditional feature selection is Pearson correlation. This algorithm also ranks features according to their correlation with our 'Class'. We can therefore decide to set a threshold and select the features or even decide to select a number of top features. But when we used threshold, we have a large number of features. We experimented with threshold of 0.3, 0.5, 0.7, 0.9, but all these thresholds gave a number of features higher than 15. So we decided to select a desired number according to their best correlation with the target ("Class"). We selected 5, 10 and 15 first features to train the models.

Moreover, we applied some bio-inspired starting by Genetic Algorithm (GA). We considered a population of size = 10, a mutation rate = 0.1, and at 5 generations; we did the Crossover of certain parents and mutation of some at each iteration and we found these features. We changed the population size to 20, then 30 and also increased the generation number, but there is no difference between the features obtained with a size of 10.

Another bio-inspired algorithm that we used is Ant Colony (AC). We used this algorithm with different numbers of ant but the best number of ants is 30, the number of iterations is 20, pheromone is initialized to 1 for all nodes. We used a K-Nearest Neighbors (KNN) classifier ($k = 5$) and this gave us the following features.

And lastly, we implemented the Bee Colony (BC) algorithm in conjunction with a k-Nearest Neighbors (KNN) classifier, utilizing $k = 5$. For the BC algorithm, we configured the number of bees to 10 and set the number of iterations to 10, while identifying the top 3 best sites for feature selection. As a result of this configuration, we achieved a reduced number of features, demonstrating the effectiveness of the BC algorithm in streamlining feature selection. This outcome reinforces the potential of bio-inspired methods to enhance the efficiency of our models while still maintaining a focus on accuracy and performance.

Bio-inspired algorithms consistently yield a reduced number of features compared to some traditional methods, which positions them favorably in the landscape of feature selection techniques. However, it is premature to declare these bio-inspired methods as definitively superior. The true measure of their effectiveness lies in the results produced by the selected features. Ultimately, it will be the accuracy and performance achieved through these features that will determine whether bio-inspired approaches genuinely enhance predictive capabilities compared to classical methods. Therefore, further evaluation and comparison are essential to draw meaningful conclusions about their overall efficacy.

The results of this experiment on feature selection algorithms are presented in Table 2. This table includes several key elements, such as the hyperparameters used, the number of features selected, and a list of those features. Each algorithm was tested to evaluate its effectiveness in selecting the best features. The data provided will allow for an in-depth analysis of the performance of each method. Thus, this table serves as an essential reference for understanding the obtained results.

As summary from results in Table 2, the ant colony algorithm is an efficient method that allows finding a very small feature subset. Other bio-inspired methods, as well as some traditional algorithms such as sequential feature selection and recursive feature elimination, also manage to reduce the number of features. On the other hand, techniques such as mutual information and Pearson correlation fail to reduce the size of the feature subset.

However, there is a need to investigate whether these feature subsets, although reduced, improve the accuracy of our models during training.

4.3 Training

We trained several ensemble models, including AdaBoost, Bagging, Gradient Boosting, and Stacking, utilizing a dataset partitioned into two subsets: 70% for training and 30% for testing. The performance of these models was assessed using multiple evaluation metrics, including accuracy, F1-score, recall, and area under the ROC curve (AUC-ROC). Among the tested models, the stacking algorithm consistently achieved superior performance across all metrics. This outcome indicates that the stacking approach, which integrates predictions from multiple base learners (Logistic Regression, K-Nearest Neighbors, Decision Tree, Support Vector Machine, and Naive Bayes), offers enhanced predictive accuracy and robustness compared to the other ensemble methods.

Table 3 shows results obtained for the stacking model.

The stacking model emerged as the most effective among all the evaluated algorithms. Across all feature selection methods, stacking consistently delivered superior performance metrics. Notably, the recall metric achieved a perfect value of 1 when using the Mutual Information method with 10 and 15 features, a distinction not observed in any other models, highlighting its exceptional performance. Furthermore, stacking achieved the highest accuracy, particularly when using features selected by the Bee Colony algorithm, which required only 16 features to deliver outstanding results. When evaluating accuracy, bio-inspired methods demonstrated clear superiority over traditional approaches. Bio-inspired algorithms directly identify features deemed critical for the task, offering a more context-aware and efficient solution. In contrast, traditional algorithms focus on the relationships, such as correlations or informativeness, between features and the target variable. For instance, the Sequential Forward Selection (SFS) method selected 21 features and achieved an accuracy of 0.999572, whereas the Bee Colony method achieved a slightly higher accuracy of 0.999648 with only 16 features. However, the Ant Colony algorithm exhibited relatively poor accuracy, attributable to its selection of only four features. This limitation suggests that the algorithm overlooked some essential features, such

Table 2 Subsets of features selected by each algorithm

Method	Hyper parameter	Number of features	List of features
Genetic algorithm	Population of size = 10, a mutation rate = 0.1, and at 5 generations	18	$V_4, V_5, V_6, V_9, V_{10}, V_{13}, V_{15}, V_{16}, V_{17}, V_{18}, V_{19}, V_{20}, V_{21}, V_{22}, V_{23}, V_{24}, V_{27}, V_{28}$
Bee colony	$k = 5$, number of bees = 10, number of iterations is 10, number of best sites = 3	16	$V_1, V_2, V_3, V_9, V_{11}, V_{13}, V_{14}, V_{15}, V_{17}, V_{18}, V_{19}, V_{20}, V_{21}, V_{23}, V_{24}, V_{25}$
Ant colony	Number of ant equal to 30, the number of iterations is 20, pheromone is initialized to 1	4	$V_{14}, V_{16}, V_{22}, V_{26}$
SFS	Logistic regression(accuracy 0.9)	21	$V_{14}, V_4, V_{12}, V_{11}, V_1, V_8, V_{18}, V_{17}, V_{24}, V_9, V_7, V_3, V_{16}, V_{28}, V_2, V_{26}, V_{22}, V_{25}, V_{23}, V_{20}$
RFE	$K = 5$	5	$V_4, V_8, V_{10}, V_{12}, V_{14}$
RFE	$K = 10$	10	$V_3, V_4, V_7, V_8, V_{10}, V_{11}, V_{12}, V_{14}, V_{16}, V_{17}$
RFE	$K = 15$	15	$V_1, V_3, V_4, V_7, V_8, V_9, V_{10}, V_{11}, V_{12}, V_{14}, V_{16}, V_{17}, V_{18}, V_{22}, V_{23}$
Mutual information	Order the mutual information values(largest to smallest)	28	$V_{14}, V_{17}, V_{10}, V_{12}, V_4, V_{11}, V_3, V_{16}, V_7, V_2, V_9, V_{21}, V_{27}, V_{18}, V_1, V_6, V_{28}, V_5, V_8, V_{19}, V_{20}, V_{24}, V_{23}, V_{26}, V_{25}, V_{22}, V_{15}, V_{13}$
Pearson's correalation	Order the correlation values(largest to smallest)	28	$V_{14}, V_{12}, V_4, V_{11}, V_3, V_{10}, V_9, V_{16}, V_1, V_2, V_7, V_{17}, V_6, V_{18}, V_5, V_{19}, V_{27}, V_{20}, V_8, V_{24}, V_{21}, V_{28}, V_{13}, V_{26}, V_{25}, V_{15}, V_{22}, V_{23}$

Table 3 Stacking model (metric)

Method	Number of features	Accuracy	F1 score	Recall	AUC-ROC
Genetic algorithm	18	0.999443	0.999444	0.999801	0.999442
Bee colony	16	0.999648	0.999649	0.999930	0.999648
Ant colony	4	0.993308	0.993345	0.996823	0.993301
SFS	21	0.999572	0.999573	0.999941	0.999571
Pearson's correlation	5	0.996934	0.996947	0.999450	0.996930
Pearson's correlation	10	0.999097	0.999100	0.999977	0.999096
Pearson's correlation	15	0.999396	0.999398	0.999988	0.999395
Mutual information	5	0.995920	0.995937	0.998478	0.995916
Mutual information	10	0.999138	0.999140	1.000000	0.999137
Mutual information	15	0.999490	0.999491	1.000000	0.999489
RFE	10	0.998353	0.998358	0.999977	0.998350
RFE	15	0.998945	0.998948	0.999988	0.998943
RFE	5	0.995920	0.995920	0.998584	0.995915

as those identified by the Bee Colony method, which significantly contributed to improved model performance. These findings underscore the importance of selecting the most relevant features and highlight the comparative advantages of bio-inspired approaches in feature selection for stacking-based models.

The second evaluated model was AdaBoost and results are presented in Table 4 and Fig. 5. Traditional feature selection algorithms demonstrate superior performance compared to bio-inspired methods when utilizing a reduced feature set. Specifically, with only five features, traditional algorithms achieve higher accuracy than any of the bio-inspired approaches.

Our analysis revealed that the Genetic Algorithm and Bee Colony methods, which utilize 16 and 18 features respectively for training with the AdaBoost model, exhibit relatively lower accuracy compared to traditional feature selection methods. While these bio-inspired algorithms are rooted in biological mechanisms, their effectiveness is not guaranteed across all contexts. Specifically, for fraud detection using AdaBoost, bio-inspired methods tend to select features that are less relevant, leading to suboptimal performance. Consequently, the application of biologically-inspired techniques may not be well-suited for fraud detection tasks involving AdaBoost.

In contrast, the performance of the third model, Bagging, was evaluated across various feature selection algorithms. As shown in Fig. 6 and Table 5, the results

Table 4 AdaBoost model (metric)

Method	Number of features	Accuracy	F1 score	Recall	AUC
Genetic algorithm	18	0.968093	0.967730	0.955220	0.968115
Bee colony	16	0.960792	0.960128	0.942274	0.960829
Ant colony	4	0.954496	0.953860	0.938851	0.968115
SFS	21	0.968691	0.968390	0.957537	0.968710
RFE	5	0.993118	0.993144	0.995154	0.993114
RFE	10	0.997040	0.997048	0.998116	0.997038
RFE	15	0.997186	0.997194	0.998338	0.997184
Pearson's correlation 5	5	0.993733	0.993757	0.995740	0.993730
Pearson's correlation	10	0.996623	0.996634	0.997940	0.996621
Pearson's correlation	15	0.997520	0.997520	0.998654	0.997518
Mutual information	5	0.993270	0.993293	0.994885	0.993268
Mutual information	10	0.997186	0.997194	0.998408	0.997184
Mutual information	15	0.997620	0.997626	0.998490	0.997619

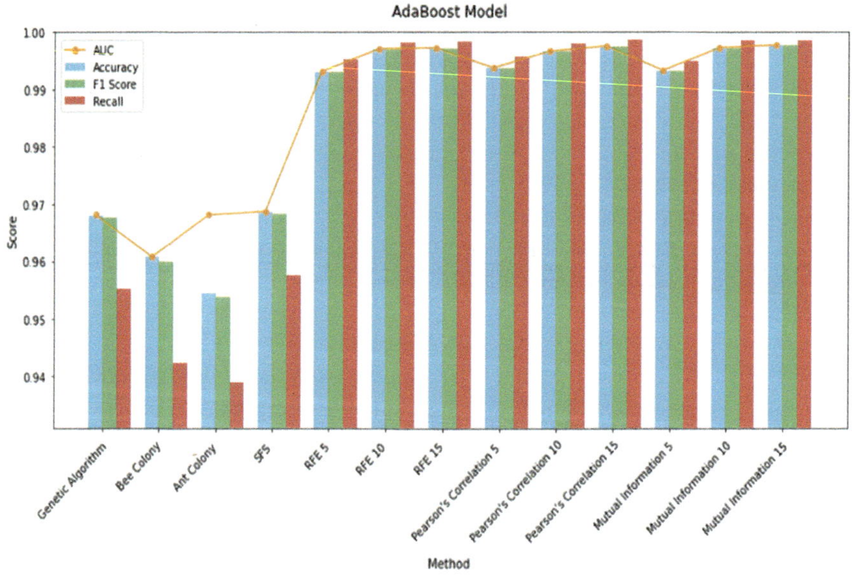

Fig. 5 All the results for AdaBoost model as histogram

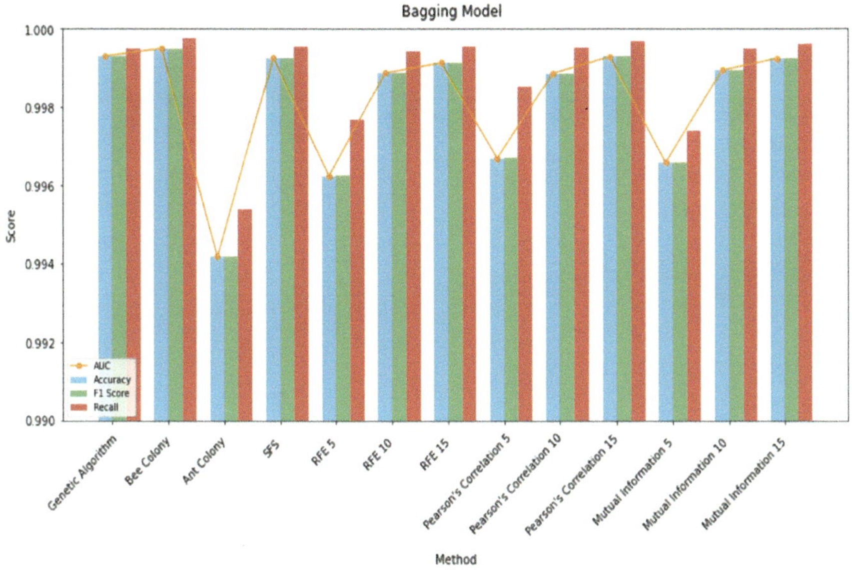

Fig. 6 Complete results for Bagging model as histogram

Table 5 Bagging models (metric)

Method	Number of features	Accuracy	F1 score	Recall	AUC
Genetic algorithm	18	0.999302	0.999304	0.999485	0.999302
Bee colony	16	0.999499	0.999500	0.999754	0.999498
Ant colony	4	0.994179	0.994179	0.995384	0.994177
SFS	21	0.999256	0.999257	0.999532	0.999255
RFE 5	5	0.996248	0.996260	0.997683	0.996246
RFE	10	0.998874	0.998877	0.999426	0.998874
RFE	15	0.999127	0.999128	0.999544	0.999126
Pearson's correlation	5	0.996694	0.996705	0.998525	0.996691
Pearson's correlation	10	0.998863	0.998865	0.999520	0.998862
Pearson's correlation	15	0.999291	0.999292	0.999684	0.999290
Mutual information	5	0.996582	0.996591	0.997402	0.996581
Mutual information	10	0.998945	0.998947	0.999497	0.998944
Mutual information	15	0.999244	0.999245	0.999614	0.999243

Table 6 Boosting model (metric)

Method	Number of features	Accuracy	F1 score	Recall	AUC
Genetic algorithm	18	0.966733	0.966167	0.948397	0.966764
Bee colony	16	0.958708	0.957681	0.932586	0.958760
Ant colony	4	0.958479	0.957734	0.938992	0.958518
SFS	21	0.967149	0.966619	0.949964	0.967179
RFE 5	5	0.961955	0.961342	0.944476	0.961985
RFE 10	10	0.964998	0.964413	0.946945	0.965029
RFE 15	15	0.965883	0.965332	0.948373	0.965913
Pearson's correlation 5	5	0.962782	0.962170	0.944991	0.962812
Pearson's correlation 10	10	0.998326	0.220588	0.131579	0.565734
Pearson's correlation 15	15	0.966252	0.965671	0.947718	0.966284
Mutual information 5	5	0.959933	0.959191	0.940145	0.959967
Mutual information 10	10	0.964746	0.964117	0.945611	0.964778
Mutual information 15	15	0.965045	0.964441	0.946554	0.965076

challenge the notion that traditional feature selection methods always outperform bio-inspired approaches. Notably, the Bee Colony algorithm achieved an accuracy of 0.999499, which surpasses all other methods. While Mutual Information, Pearson Correlation, and Recursive Feature Elimination demonstrated slightly lower accuracy values with 15 features, the number of features does not appear to be the sole determinant of performance. For example, the Genetic Algorithm and SFS methods, utilizing 18 and 21 features respectively, did not outperform Bee Colony, which achieved superior accuracy with only 16 features. This suggests that the relevance of selected features plays a more critical role than their quantity in influencing model performance.

The last model is gradient boosting and the results obtained are presented in Table 6 and Fig. 7.

Among the evaluated models, one exhibited notable weaknesses in certain metrics. Specifically, its recall value was 0.131579, a result not observed in any other models. However, despite this limitation, the model achieved an accuracy of 0.998326, which remained higher than that of the others. This performance, coupled with the use of

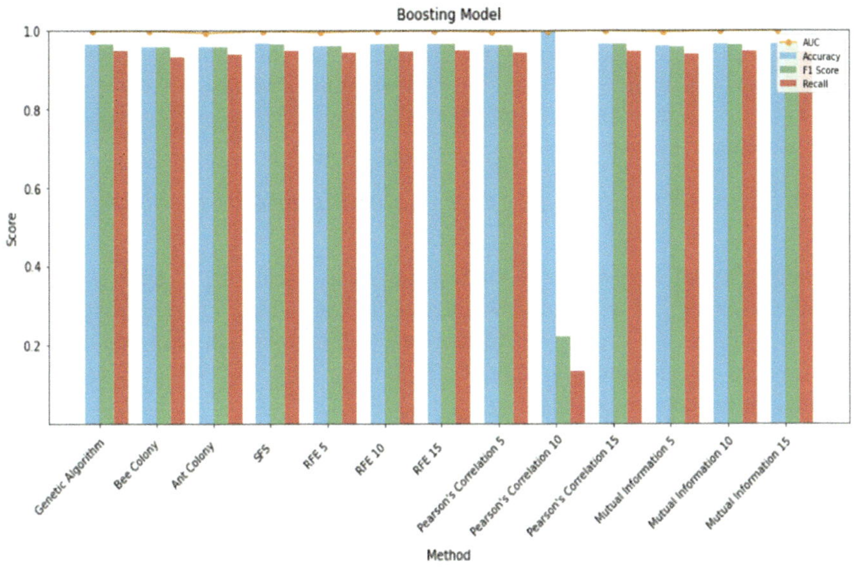

Fig. 7 Complete results for gradient boosting

only 10 features, reinforces the observation that the number of features alone is not sufficient to enhance model accuracy.

The observed performance variations among the stacking, AdaBoost, Bagging, and Boosting models when trained on the same feature sets can be attributed to several factors. For instance, AdaBoost prioritizes misclassified samples in each iteration, while Bagging leverages bootstrap sampling to create diverse subsets. Stacking employs five base models (Logistic Regression, K-Nearest Neighbors, Decision Tree, Support Vector Machine, and Naive Bayes), which introduces additional variability due to its unique mechanism of integrating predictions from diverse algorithms. These differences in operational strategies explain the performance discrepancies, even when identical feature sets are used.

Moreover, the optimization of hyperparameters plays a critical role in determining model performance. Each model requires specific hyperparameter configurations, and suboptimal tuning can lead to reduced performance, as observed in the case of Boosting. This underscores the importance of carefully selecting and fine-tuning hyperparameters to maximize the potential of each model within the context of the selected features.

In summary, the amount of training data significantly influenced the performance of the models. Certain algorithms demonstrate superior performance with larger datasets, while others are more effective with smaller data volumes. For instance, Boosting algorithms typically underperform when a large number of estimators and iterations are used in conjunction with extensive datasets. In our study, the Boosting model was configured with 100 estimators and 100 iterations, combined with a substantial dataset size. This configuration likely contributed to its comparatively lower performance, underscoring the importance of balancing model parameters with the size and characteristics of the data to optimize results.

4.4 Choose Traditional Feature Selection or Bio-Inspired

A comparison of our work with similar works has been done in Table 7 which is adapted from Table 1 with the same columns and one added line refers to our work.

The findings of our study elucidate a nuanced interplay between traditional and bio-inspired feature selection methods, highlighting their respective strengths and limitations in various contexts.

Sequential forward selection (SFS), a traditional method, demonstrates a compelling ability to deliver superior accuracy in numerous scenarios. By systematically starting with a minimal feature set and progressively adding features, SFS effectively navigates the feature space. Its notable performance underscores its potential as an efficient choice for feature selection, particularly in applications where accuracy is paramount.

Conversely, bio-inspired methods, such as the bee colony algorithm, exhibit exceptional performance, surpassing SFS in specific instances. This superior accuracy can be attributed to the bee colony method's inherent ability to explore the feature space efficiently and adapt to data variability, emphasizing the unique advantages of nature-inspired approaches in tackling complex feature selection tasks.

However, not all bio-inspired techniques demonstrate similar efficacy. For instance, the ant colony algorithm, despite its biological underpinnings, showed subpar accuracy compared to other evaluated methods. This underperformance underscores critical limitations in its adaptability to certain scenarios and emphasizes the need for further refinement to enhance its competitiveness.

As shown in Table 7, this study is contextualized within the broader body of research on feature selection for fraud detection. This comparative analysis is vital for situating our contribution within the existing landscape, enabling us to assess the strengths and limitations of our approach. Numerous researchers have employed a variety of methods tailored to the specificities of their studies. While some have relied on traditional statistical-based techniques, which have consistently demonstrated their effectiveness, others have embraced bio-inspired methods, reflecting a diversity of perspectives and approaches in the field. Each strategy offers unique value, underscoring the need for a comprehensive comparison to position our work within this dynamic framework.

By examining the outcomes of these methodologies, the goal is to justify our selection of techniques and uncover opportunities for further refinement. Unlike these studies that primarily apply feature selection algorithms to identify features, train models, and evaluate performance based on accuracy, our research took a more nuanced approach. We focused on determining the optimal number of features for precise fraud detection while exploring the conditions under which bio-inspired or

Table 7 Comparison of our work and other works

Authors	Aproaches	Contributions	Conditions	Dataset
Jain and Singh [2]	Firefly algorithm	Improve the accuracy to 96.23%	NO	Australian credit data
Authors [9, 17, 35]	Fruit fly algorithm (FFA)	Increase the accuracy	NO	NSL-KDD datasets
Jain and Singh [1]	Filter and Wrapper	Improve the accuracy	NO	German credit dataset
Priscilla and Prabha [31]	Utual Information (MI) and Recursive Feature Elimination (RFE)	The accuracy was reduced to 84.8%	NO	dataset from the IEEE-CIS
Poongodi and Kumar [29]	Information Gain	improve the accuracy to 94.10%	NO	Credit card clients dataset
Moral-Rubio et al. [13]	Informative Variable Identifier (IVI)	Increase the accuracy 76%	NO	Malware datasets, phishing website URLs, credit card fraud transaction datasets
Pandurangan [27]	Metaheuristic feature selection	Increase the accuracy 94.78%	NO	German credit dataset and credit card fraud detection dataset
Mienye and Sun [21]	Information Gain(IG) and Genetic Algorithm (GA)	Sensitivity was 0.997 and specificity was 0.994	NO	European cardholders dataset
Prabhakaran et al. [30]	Oppositional Cat Swarm Optimization- based(OCSO)	Increase the accuracy 99.97 %	NO	Benchmark dataset
Omoworare [24]	Correlation matrix	Increase the accuracy 98 %	NO	Dataset created
Gupta [11]	Grasshopper algorithm	Increase accuarcy to 95.49%	NO	European credit-cardholders
Our study	Genetic Algorithm, Ant Colony, Bee Colony, Sequential Feature Selection, Recursif Feature Elimination, Mutual Information, Pearson's Correlation	Select directly optimal (k) number of feature(Bio-inspired), put the criteria to select (k) features(traditional) to have high accuracy	$\begin{cases} \text{if } k < 10 & \text{Bio-inspired} \\ \text{else} & \text{Traditional} \end{cases}$	European cardholders https://www.kaggle.com/datasets/nelgiriyewithana/credit-card-fraud-detection-dataset-2023 dataset

traditional algorithms are most effective. For instance, $k = 10$ represents the threshold from which the decision can be made. This parameter can be optimized in further studies. This distinction enriches our understanding of the challenges inherent to fraud detection and informs our future research directions, emphasizing the practical application of feature selection methods under varying scenarios.

These observations suggest that while bio-inspired methods hold significant promise, their performance varies considerably, necessitating context-specific evaluation and optimization. Our results underscore the importance of selecting a feature selection method tailored to the objectives and characteristics of the analysis. Both traditional and bio-inspired approaches offer distinct advantages, and their relative effectiveness depends heavily on the dataset and experimental conditions.

5 Conclusion

The goal of this work was to investigate traditional and bio-inspired feature selection techniques in regards to the credit card fraud detection.

To achieve this goal, a dataset that contains information about credit card transactions was used. Four traditional feature selection techniques i.e. SFS, RFE, MI and Pearson Correlation, and three bio-inspired feature selection techniques including Genetic Algorithm (GA), Ant Colony (AC) and Bee Colony (BC) selection methods were implemented to find the number of relevant features. Four ensemble learning models have been trained on the selected features to evaluate the performance. Results revealed the subsets of features that are suitable for fraud detection in both categories of feature selection and number of features that indicate the choice of feature selection category.

Future work includes studying combination of bio-inspired algorithms and combination between bio-inspired and traditional to select features offering precise fraud detection. Additionally, methods like Principal Component Analysis (PCA), t-distributed Stochastic Neighbor Embedding (t-SNE) will be studied concerning detection of fraud.

Acknowledgements We extend our sincere gratitude to AIMS Cameroon for offering effective training and agreeable stay to Fidele, that permitted him to acquire requisite background about data science to conduct this research.

References

1. Anurag Jain Ajeet Singh. Adaptive credit card fraud detection techniques based on feature selection method. *Applied Science*, pages 167–178, 2019.
2. Anurag Jain Ajeet Singh. Hybrid bio-inspired model for fraud detection with correlation based feature selection. *Taru Publications*, 2021.
3. Gustavo E. Vazquez Ignacio Ponzoni Axel J. Soto, Rocío L. Cecchini. An evolutionary approach for feature selection applied to admet prediction. *Inteligencia Artificial. Revista Iberoamericana de Inteligencia Artificial*, 2008.
4. Mouna Tarik Ayoub Mniai and Khalid Jebari. A novel framework for credit card fraud detection. *International Conference on Integrated Intelligence and Communication Systems (ICIICS)*, 11, 2023.
5. Peter Buhlmann. Bagging, boosting and ensemble methods. *ETH Zurich, Seminar fur Statistik*, 2012.
6. Danny Butvinik. Feature selection-exhaustive overview. *Analytics Vidhya*, 2021.
7. Hassan Chouaib. *Selection of features: methods and applications*. PhD thesis, 2011.
8. Andrew Engel. Categorical variables for machine learning algorithms. *Towards Data Science*, 2022.
9. Ali Kadhum Idrees Faezah Hamad Almasoudy, Wathiq Laftah Al-Yaseen. Differential evolution wrapper feature selection for intrusion detection system. *MDPI*, 167:1230–1239, 2020.
10. CAMPOS Guillermo. *An Improved Artificial Bee Colony for Feature Selection in QSAR*. PhD thesis, 2021.
11. Parveen Gupta, Manu Jyoti; Sehgal. Optimizing credit card fraud detection: Classifier performance and feature selection empowered by grasshopper algorithm. *International Journal of Performability Engineering*, 20:p177, 2024.
12. Ritesh Gupta. Accuracy, precision, recall, f-1 score, confusion matrix, and auc-roc. *Medium*, 2023.
13. Santiago Moral-Rubio Sergio Munoz-Romero Jacobo Chaquet-Ulldemolins, Francisco-Javier Gimeno-Blanes and Jose-Luis Rojo-Alvarez. On the black-box challenge for fraud detection using machine learning (i): Linear models and informative feature selection. *Applied Science*, 3328:076–3417, 2022.
14. Kewei Cheng-Fred Morstatter Jundong Li, Suhang Wang. Feature selection: A data perspective. *ACM Computing Surveys*, 2016.
15. Roman Zajdel MaciejKusy. A weighted wrappers approach to feature selection. *ACM Computing Surveys*, 31:685–696, 2021.
16. Raneem Qaddoura; Hossam Faris Mariam M. Biltawi. Optimizing feature selection and oversampling using metaheuristic algorithms for binary fraud detection classification. *Artificial Intelligence Applications and Innovations*, 20:452–462, 2023.
17. Farhad Soleimanian Gharehchopogh Payam Alemi Maryam Samadi Bonab, Ali Ghaffari. A wrapper-based feature selection for improving performance of intrusion detection systems. *Wiley Online Library*, 33, 2020.
18. Abel Efe Edje Clive Asuai Ufuoma John Osame Maureen Ifeanyi Akazue, Irene Alamarefa Debekeme*. Unmasking fraudsters: Ensemble features selection to enhance random forest fraud detection. *Journal of Computing Theories and Applications*, 3, 2023.
19. Nad Al Sheba Meydan Freezone, Meydan Road. Prevention de la fraude par carte de credit proteger votre entreprise strategies de prevention de la fraude par carte de credit, 2024. Accessed: April 14, 2024.
20. Voica Radescu Robert Loredo Kirsten Voigt Constantin von Altrock Andreas Rudnik Michele Grossi, Noelle Ibrahim. Mixed quantum-classical method for fraud detection with quantum feature selection. *IEEE Transactions on Quantum Engineering*, 3, 2022.
21. Ibomoiye Domor Mienye and Yanxia Sun. *A Machine Learning Method with Hybrid Feature Selection for Improved Credit Card Fraud Detection*. PhD thesis, University of Johannesburg, South Africa, 2023.

22. Dr. G.Dheepa N.Geetha. A hybrid deep learning and modified butterfly optimization based feature selection for transaction credit card fraud detection. *Journal of Positive School Psychology*, 2022.
23. Stig-Helmer Olsson. An introduction to tomatoes. Master's thesis, Chalmers University of Technology, Sweden, 2010.
24. Boluwatife Joseph Omoworare. Feature selection and machine learning algorithms for an improved credit card fraud detection system. https://norma.ncirl.ie/6244/, 2023. Accessed: April 2, 2024.
25. R Opitz, D.; Maclin. Popular ensemble methods: An empirical study. *Journal of Artificial Intelligence Research*, 1999.
26. D. S. Rao P. Swetha. Effective feature selection-based meta-heuristics optimization approach for spam detection. *SN Computer Science*, 11:734–745, 2023.
27. P. Rajesh Pandurangan. Design of metaheuristic feature selection with deep learning based credit card fraud detection model, 2022.
28. Tin H. Pham and Bijan Raahemi. Bio-inspired feature selection algorithms with their applications: A systematic literature review. *IEEE Access*, 11:43733–43758, 2023.
29. Kannan Poongodi and Dhananjay Kumar. Support vector machine with information gain based classification for credit card fraud detection system. *The International Arab Journal of Information Technology*, 18, 2021.
30. N. Prabhakaran and R. Nedunchelian. Oppositional cat swarm optimization-based feature selection approach for credit card fraud detection. *Computational Intelligence and Neuroscience*, 2023:13, 2023.
31. C. Victoria Priscilla and D. Padma Prabha. A two-phase feature selection technique using mutual information and xgb-rfe for credit card fraud detection. *International Journal of Advanced Technology and Engineering Exploration*, 8(85), 2021.
32. Matt Rej. Credit card fraud statistics (2024). *Merchant Cost Consulting*, 2023.
33. Sabeena S. Optimal feature subset selection using ant colony optimization, 2015.
34. Ibomoiye Domor Mienye; Yanxia Sun. A survey of ensemble learning: Concepts, algorithms, applications, and prospects. *IEEE Xplore*, 2022.
35. Faezah Hamad Almasoudy Wathiq Laftah Al-Yaseen, Ali Kadhum Idrees. Wrapper feature selection method based differential evolution and extreme learning machine for intrusion detection system. *MDPI*, 132, 2022.

Leveraging Deep Learning for Enhanced Software Fault Prediction Using Error-Type Metrics

Khoa Phung, Emmanuel Ogunshile, and Mehmet E. Aydin

Abstract Software Fault Prediction (SFP) is a crucial strategy in software quality assurance, aiming to reduce costs and optimise resource allocation by identifying potentially faulty software modules based on key project characteristics. Despite significant advancements, SFP methods often reach a 'performance ceiling' due to the constraints imposed by the limited size of datasets available from public repositories and the difficulties in selecting optimal software metrics for specific application domains. Furthermore, while traditional machine learning approaches have predominantly been used to predict fault proneness, the possibilities offered by more sophisticated techniques like Deep Neural Networks (DNNs) remain underutilised. This study introduces a novel application of DNNs, trained using Error-type Metrics, to predict faults in open-source software projects. These metrics are application-independent and have been shown to enhance predictive accuracy by enriching the training data with a broader range of information, thus helping to break through the existing performance limitations. Our empirical findings indicate that models trained with Error-type Metrics significantly outperform those using traditional CK metrics, achieving improvements of up to 40% in AUC and ROC scores. Our method outstrips even the latest DNN models incorporating advanced self-attention mechanisms, achieving performance gains of up to 17.86%.

K. Phung (✉) · E. Ogunshile · M. E. Aydin
School of Computing and Creative Technologies, University of the West of England, Bristol, UK
e-mail: khoa.phung@uwe.ac.uk

E. Ogunshile
e-mail: emmanuel.ogunshile@uwe.ac.uk

M. E. Aydin
e-mail: mehmet.aydin@uwe.ac.uk

N. Polatidis et al. (eds.), *Intelligent Distributed Computing XVII*, Studies in Computational Intelligence 1203, https://doi.org/10.1007/978-3-031-87639-4_10

1 Introduction

Software Fault Prediction (SFP) is a technique that classifies whether a software module is faulty by considering underlying characteristics collected from software projects [1]. SFP can be accomplished by dividing the modules/classes of a software system into fault-prone and not-fault-prone categories. The fault datasets, which include software metrics, obtained from previous releases of the same project or projects with similar domains are used to train machine learning models [2–4]. The trained models are subsequently applied to the current software project's modules, classes or methods to classify them as fault-prone or not fault-prone.

SFP has achieved significant advancements in the past decade. However, a prominent limitation associated with these advancements, termed the 'performance ceiling' [5, 6], has sparked debates within the research community. It has been observed that enhancing the machine learning techniques employed does not necessarily lead to improved SFP performance, primarily because the small-scale datasets used for these studies are derived from public repositories like NASA [7] and PROMISE [8]. Further, the existing literature has inadequately addressed the challenge of selecting the most appropriate software metrics for fault prediction [9]. This has led to a stagnation in SFP advancements.

Also, a major gap in the current SFP studies is the limited exploration of deep learning techniques. Deep learning, specific techniques such as Deep Neural Networks (DNNs), holds tremendous promise for the future of SFP. Unlike traditional machine learning models, DNNs can automatically learn and extract high-level representations from the data, thereby capturing complex patterns and improving model accuracy [10]. These capabilities of DNNs indicate a significant untapped potential in the field of SFP, an aspect that this study aims to explore.

Motivated by these challenges, this paper presents a DNN-based approach for fault-proneness prediction, trained using Error-type Metrics. These metrics, initially developed in [11] and officially introduced in [12], can be applied across various application domains and have demonstrated their ability to enhance the performance of machine learning models in fault-proneness prediction. To the best of our knowledge, this is the first study to utilise Error-type Metrics for constructing a DNN-based model for fault-proneness prediction.

Our study aims to provide empirical evidence to address the following research questions:

- **RQ1: To what extent can the use of Error-type Metrics in training DNNs with open-source software projects enhance the 'performance ceiling' compared to state-of-the-art machine learning and deep learning techniques?**–This research question aims to compare the performance of DNN-based fault prediction models with state-of-the-art machine learning and deep learning approaches when Error-type Metrics are used as independent variables.
- **RQ2: Are Error-type Metrics more effective predictors of faults compared to CK metrics when used to train DNN models?**–This research question aims

to empirically compare the performance differences between DNN models constructed using Error-type Metrics and those constructed using CK metrics [13]. The rationale for using CK metrics as a benchmark for comparison is detailed in Sect. 3.2.

In contrast to the majority of studies which utilise publicly available datasets from repositories such as NASA and PROMISE, this study uses metrics extracted from four open-source software projects, as detailed in Sect. 3.1. The fault data for these projects was obtained from the BugHunter Dataset [14].

The rest of this paper is organised as follows. Section 2 provides a review of related work. Section 3 describes the methodology used. Section 4 outlines the experimental results and discusses their implications. Finally, Sect. 5 concludes the paper and provides future research directions.

2 Related Work

This section reviews the literature on software metrics used in SFP as well as the application of deep learning in SFP.

2.1 Software Metrics

Software metrics, as defined by Goodman [15], represent different properties of a software product that are continually monitored throughout the development lifecycle. These properties, including aspects like coupling, cohesion, and inheritance, each correspond to a specific metric. Such metrics are crucial for determining the external quality of software modules in terms of factors like testability, reliability, and fault-proneness [16].

Over the past three decades, a wide range of software metrics has been proposed. These include object-oriented metrics such as CK metrics suite [13], MOODS metrics suite [17], Bansiya metrics suite [16], and traditional metrics like size metrics (e.g., Function Points–FP, SLOC, KSLOC), quality metrics (e.g., Defects per FP after delivery, Defects per SLOC or KSLOC after delivery), system complex metrics [18], and Halstead metrics [19]. As highlighted by Rathore et al. [2], the performance of metrics can vary based on the environment (e.g., open-source vs. commercial).

In our previous work [11], we proposed a Stream X-Machine approach [20] to model and generate test cases for different types of Java runtime errors. These test cases were then used to extract error-type data (referred to as *Error Specification Machine* values or *ESM values*) from the source code. In a subsequent study [12], these ESM values were evaluated as a combined objective, resulting in a novel set of software metrics called *Error-type software metrics* (or Error-type Metrics for short). As these metrics are based on error-type models, they can be applied to any application domain, potentially addressing the challenge of selecting the best combination of metrics for each different application domain in a given context.

2.2 Deep Learning in SFP

Kumudha et al. [21] suggested an SFP method which employed a conventional neural network called radial basis function neural network (RBFNN) along with a new model called adaptive dimensional bio-geography based optimisation–ADBBO. To evaluate the effectiveness of this approach, the ADBBO-based RBFNN model was applied to five datasets obtained from the NASA repository. The authors compared the proposed model with the previously established algorithms in the literature using the same five NASA datasets. The empirical results showed that the ADBBO-RBFNN model outperformed the existing approaches in fault-proneness prediction.

Yang et al. [22] proposed a novel SFP approach, called Deeper, which utilised deep learning techniques to generate expressive features from a set of initial change features via a deep belief network algorithm. The effectiveness of the proposed approach was evaluated on six datasets derived from large open-source projects including Bugzilla, Columba, JDT, Platform, Mozilla, and PostgreSQL. The findings demonstrated that the Deeper model identified 32.22% more bugs than Kamei et al. [23]'s approach (51.04% versus 18.82% on average). Furthermore, Deeper achieved statistically significantly higher F1-scores of 0.22–0.63 than Kamei et al. [23]'s approach on four out of the six projects.

Manjula et al. [24] proposed a novel approach for early fault prediction in software development by combining genetic algorithm (GA) for feature optimisation and DNN for classification. The proposed approach was evaluated using four datasets from the PROMISE repository. The empirical results indicated the superior performance of this approach over other existing techniques, with accuracy rates of 97.82% for the KC1 dataset, 97.59% for the CM1 dataset, 97.96% for the PC3 dataset, and 98.00% for the PC4 dataset.

Recently, Yu et al. [25] proposed a deep learning model named Defect Prediction via Self-Attention mechanism (DPSAM). The model leverages a self-attention mechanism to extract semantic features of programs and predict defects. DPSAM achieved an average F1-Score of 70% in both within-project and cross-project defect prediction on seven open-source projects of the PROMISE dataset.

In contrast to the surveyed literature, our study distinguishes itself in two key aspects. Firstly, instead of using conventional or object-oriented metrics, we harness the recently proposed *Error-type software metrics* in training DNN models. The unique advantage of Error-type Metrics is their ability to provide detailed insights into the fault-proneness of software modules, given their derivation from an extensive catalogue of Java runtime errors. Secondly, while the majority of existing studies utilise public datasets from repositories such as NASA or PROMISE, we extract software metrics from open-source software projects that are still actively maintained. This approach allows us to train our models on datasets that potentially reflect a more realistic software development environment. Consequently, by constructing a DNN-based SFP model utilising Error-type Metrics derived from real-world software projects, our study bridges the research gap and contributes to the further enhancement of fault prediction models applicable across a wide range of domains.

3 Methodology

This section describes our methodology including the details of the software fault datasets, independent variables (software metrics), the dependent variable, the evaluation measures, and data preprocessing.

3.1 Software Fault Datasets

In our study, we utilise software fault datasets derived from the BugHunter Dataset [14]. Unlike public repositories such as NASA and PROMISE, which have been reported to contain potentially erroneous or superfluous information [26], the BugHunter Dataset focuses on the faulty and fixed states of source code irrespective of release versions, thereby capturing changes in software metrics when bugs are resolved. This approach provides a more holistic view of software faults across the project lifespan [14].

Four actively maintained, open-source Java software projects are chosen for our study. A summary of these projects is provided in Table 1.

Table 1 The selected projects and their descriptions

Project name	Forks	kLOC	Description
ANTLR v4	1559	68	A popular software in the field of language processing It is a powerful parser generator for reading, processing, and executing or translating structured text or binary files
JUnit	2826	43	A Java framework for writing and designing unit tests
OrientDB	792	621	A popular document-based NoSQL graph database Mainly famous for its speed and scalability
Elastic Search	14303	995	A popular RESTful search engine

3.2 Independent and Dependent Variables

In this study, the independent variables of the fault datasets are the four ESM values corresponding to four types of Java runtime error which are Index Out Of Bounds Exception, Null Pointer Exception, Class Cast Exception, and Arithmetic Exception.

To validate the effectiveness of Error-type Metrics, we also train DNN models on the well-established Chidamber & Kemerer (CK) metrics suite [13]. The CK metrics suite comprises six metrics, namely Coupling Between Objects (CBO), Weighted Methods per Class (WMC), Response for a Class (RFC), Lack of Cohesion in Methods (LCOM), Depth of Inheritance Tree (DIT), and Number of Children (NOC). These metrics have been proven to produce better fault prediction results compared to other metrics such as LOC and complexity metrics [27]. Also, 49% of the researchers have used CK metrics in their studies [28]. Here, they are used as a benchmark.

The dependent variable of the study is the fault proneness of software modules, which is referred to as a class in a given Java system.

3.3 Evaluation Measures

To evaluate the performance of the DNN models, we use four different types of evaluation measures including Accuracy, F1-Score, Area Under Curve (AUC), and Receiver Operating Characteristic (ROC) Analysis.

$$AUC = \frac{S_0 - n_0(n_0 + 1)/2}{n_0 * n_1} \tag{1}$$

AUC is a measure that provides an overall evaluation of performance across all potential classification thresholds. Hand et al. [29] proposed an approach illustrated in Eq. (1) for calculating the AUC of a classifier for binary classification. n_0 and n_1 are the numbers of positive and negative examples, respectively, and $S_0 = \sum r_i$, where r_i is the rank of the ith positive example in the ranked list.

ROC analysis is a statistical method used to examine and utilise the correlation between the Sensitivity and Specificity of a binary classifier. Specificity, or the true negative rate, measures the proportion of true negative instances that are correctly classified [30]. In ROC analysis, the true positive rate is typically plotted against the false positive rate, which is calculated as one minus the true negative rate.

3.4 Data Preprocessing

The majority of features in the Error-type Metrics and CK metrics of the four projects exhibit a notable degree of left skewness, which indicates that they are not normally distributed. As a result, before training the DNN models, we preprocessed the datasets using the following steps: First, we split the datasets into training (80%) and testing (20%) sets accordingly. Second, we normalise the features by using the Box-Cox algorithm, which applies a range of power transformations (rather than the traditional square root, log, and inverse) on the datasets. According to Osborne [31], Box-Cox is a powerful technique to improve the data normalisation process for both positively and negatively skewed features.

3.5 Proposed DNN Architecture

In this subsection, we propose the architecture of the DNN-based model for fault-proneness prediction. We set the number of neurons for each layer as follows: 64 neurons for the first three layers, 128 neurons for the next three layers, and 256 neurons for the last three layers.

We use the Rectified Linear Unit (ReLU) as the activation function for all layers. ReLU is employed because it could improve the learning speed of the various deep neural networks and help avoid the vanishing gradient problem [32]. We investigated different neural network model architectures on two sets of software metrics of the four software projects. We adjusted the number of layers, the number of neurons in each layer, and the activation function of the neural network several times. We changed the number of layers from one to four and added one layer to the neural network architecture each time. Finally, we discovered that the above parameter settings achieved the best performance overall.

4 Experimental Results and Discussions

This section outlines the experimental results, answers research questions outlined in Sect. 1, and discusses the implications of the study.

4.1 Experimental Results

This subsection presents the outcomes derived from the evaluation measures delineated in Sect. 3.3. These measures were utilised to gauge the performance of the DNN models constructed using Error-type Metrics and CK metrics. Following this, we address the two research questions RQ1 and RQ2.

Addressing RQ1: Across all four projects, the DNN models yielded favourable Accuracy rates: 0.789 (ANTLR v4), 0.741 (JUnit), 0.705 (OrientDB), and 0.705 (Elastic Search). Additionally, high F1-Scores were achieved: 0.881 (ANTLR v4), 0.837 (JUnit), 0.727 (OrientDB), and 0.855 (Elastic Search). Considering the JUnit

Table 2 Experimental results

Projects	Independent variables	F1-Score	Precision	Recall	Accuracy	AUC
ANTLR v4	CK Metrics	0.835	0.850	0.820	0.741	0.595
	Error-type Metrics	0.881	0.812	0.962	0.789	0.706
JUnit	CK Metrics	0.836	0.739	0.963	0.722	0.477
	Error-type Metrics	0.837	0.776	0.909	0.741	0.666
OrientDB	CK Metrics	0.708	0.715	0.701	0.684	0.747
	Error-type Metrics	0.727	0.729	0.725	0.705	0.781
Elastic Search	CK Metrics	0.810	0.803	0.817	0.693	0.563
	Error-type Metrics	0.855	0.822	0.891	0.705	0.630

project, the DNN was able to outperform contemporary machine learning algorithms presented in [12] by a margin of up to 36%. Specifically, in [12], the F1-Scores of Decision Tree, Logistic Regression, and Naïve Bayes were 0.5333, 0.6154, and 0.5882, respectively. Additionally, a comparison with contemporary deep learning techniques showed that our model demonstrated superior performance. For instance, it achieved an average F1-Score of 0.825 which is 17.86% higher than the average F1-Score of the DPSAM model proposed by Yu et al. [25]. These findings underscore the robustness of our DNN models, particularly when trained using Error-type Metrics, underscoring their efficacy in predicting software fault-proneness within industrial-scale open-source software projects.

Addressing RQ2: As observed in Table 2, the DNN models constructed with CK metrics demonstrated commendable performance on all projects with F1-Score values of 0.835 (ANTLR v4), 0.836 (JUnit), 0.708 (OrientDB), and 0.810 (Elastic Search). Comparatively, the DNN models built using Error-type Metrics outperformed the models created with CK metrics. To illustrate, regarding F1-Score, Error-type Metrics improved DNN performance by 5.6% compared to CK metrics in the Elastic Search project, while in the JUnit project, both DNN models delivered a similar performance with only a marginal 0.1% difference. In terms of Accuracy, Error-type Metrics improved DNN performance by 6% relative to CK metrics in the ANTLR v4 project. Consequently, Error-type Metrics emerge as superior predictors of faults, despite the improvements in F1-Score and Accuracy being somewhat modest (ranging between 0.1% and 6%).

However, Error-type Metrics considerably augmented the performance of the DNN models in terms of AUC. For instance, DNN models trained using Error-type Metrics outstripped those trained with CK metrics, with the AUC rates improved by

19%, 40%, and 12% in the ANTLR v4, JUnit, and Elastic Search projects, respectively. Furthermore, Figs. 1, 2, 3, and 4 depict the ROC analysis of the DNN models built with Error-type Metrics and CK metrics. From these figures, it is apparent that the DNN models constructed using Error-type Metrics generate more convex curves compared to those built with CK metrics, implying that Error-type Metrics provide DNN models with more valuable insights about faults, thus significantly enhancing their performance relative to CK metrics.

4.2 *Threats to Validity*

Whilst the results lend credence to the efficacy of Error-type Metrics in enhancing the performance of DNN models in fault-proneness prediction, there exist certain caveats which may potentially impact the study's conclusions. The foremost concern pertains to the generalisation of the findings. Our experiments were conducted on four large-scale open-source software projects. Nevertheless, additional datasets can also be subjected to experimentation. Secondly, we constructed varying DNN models to compare the utility of Error-type Metrics against CK metrics, both of which are posited to have a strong correlation with software module fault proneness. Error-type Metrics could also be compared with other software metrics that are potent predictors of faults, such as the MOODS metrics suite [17], Bansiya metrics suite [16], System complex metrics [18], Halstead metrics [19], change metrics [33], etc. Lastly, our creation and evaluation of fault prediction DNN models were based on open-source project datasets. However, software systems developed by various organisations may exhibit distinct fault patterns. As a result, one needs to take care of the underlying pattern of the software system before applying the proposed approach.

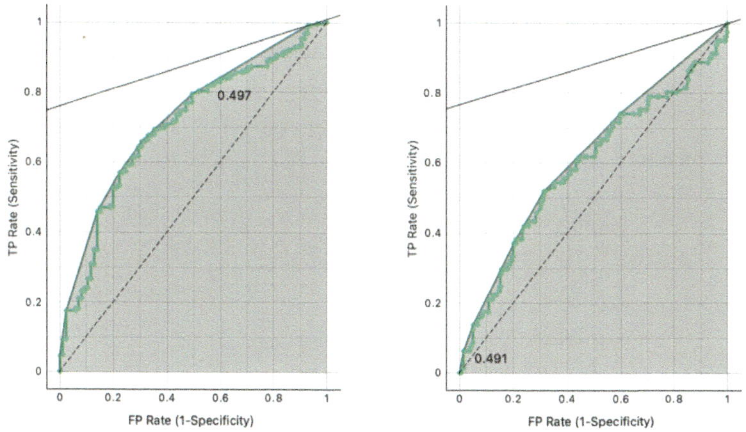

(a) Error-type Metrics (b) CK metrics

Fig. 1 ANTLR v4–ROC of DNN models built with Error-type Metrics **a** and CK metrics **b**

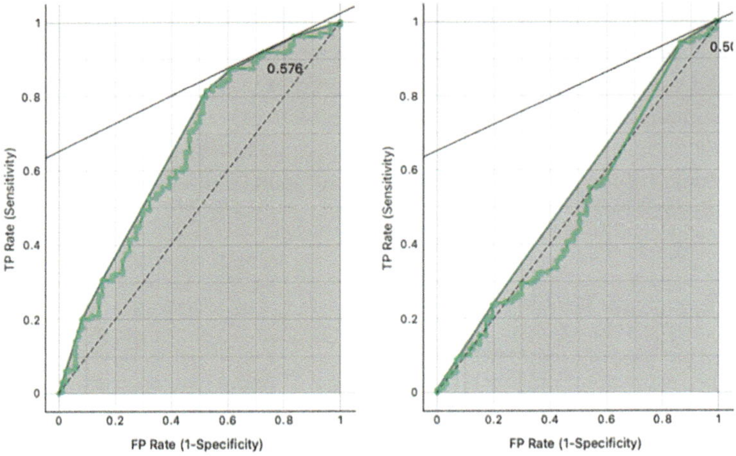

(a) Error-type Metrics (b) CK metrics

Fig. 2 JUnit–ROC of DNN models built with Error-type Metrics **a** and CK metrics **b**

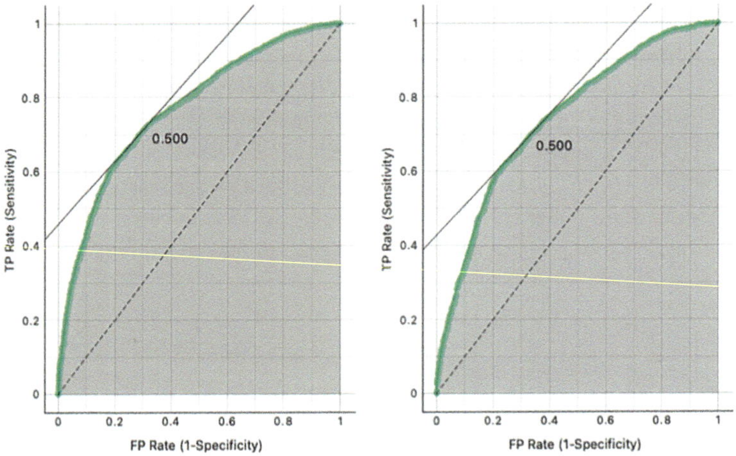

(a) Error-type Metrics (b) CK metrics

Fig. 3 OrientDB–ROC of DNN models built with Error-type Metrics **a** and CK metrics **b**

5 Conclusions

In this study, to address the gap in the current literature and industry practices, we proposed a DNN-based model trained on Error-type Metrics, diverging from the conventional use of CK metrics. Our selection of Error-type Metrics was underpinned by their potency as rich and informative predictors of software faults, rooted in their inherent correlation with error and exception handling within software systems.

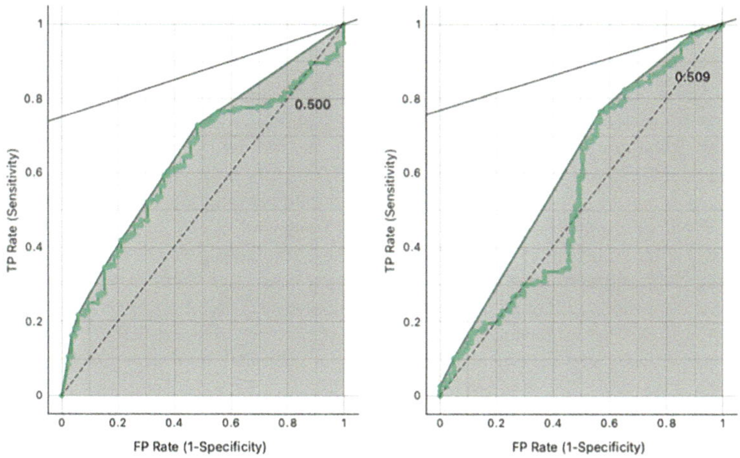

(a) Error-type Metrics (b) CK metrics

Fig. 4 Elastic Search–ROC of DNN models built with Error-type Metrics **a** and CK metrics **b**

The findings from this research, built upon experimental results, have robustly affirmed the superiority of Error-type Metrics in training high-performing DNN models. Our comparative analysis highlighted that DNN models developed using Error-type Metrics outperform those trained with CK metrics, exhibiting an enhancement of up to 40% in the AUC and ROC. Our proposed approach not only outperformed traditional machine learning techniques in [12] when trained on industrial-scale open-source software projects but also demonstrated superior results compared to state-of-the-art DNN models, including those leveraging self-attention mechanisms proposed by Yu et al. [25], with an improvement of up to 17.86%.

Thus, this study has made two significant contributions:

1. Advocated for using DNN-based models in software fault-proneness prediction, presenting Error-type Metrics as a compelling alternative to traditional CK metrics.
2. Evidenced, through experiments on large-scale open-source software projects, the ability of DNN models to outperform conventional machine learning techniques when trained with Error-type Metrics.

Future research could expand the comparative study of Error-type Metrics to encompass other software metric sets prevalent in the literature. Additionally, the experimental scope could be broadened to include more diverse open-source and industrial projects, thereby bolstering the generalisability of our findings. These avenues of exploration could help address the limitations identified in Sect. 4.2.

References

1. Ezgi Erturk and Ebru Akcapinar Sezer. "Iterative software fault prediction with a hybrid approach". In: *Applied Soft Computing* 49 (2016), pp. 1020–1033.
2. Santosh S Rathore and Sandeep Kumar. "A study on software fault prediction techniques". In: *Artificial Intelligence Review* 51.2 (2019), pp. 255–327.
3. Ruchika Malhotra. "A systematic review of machine learning techniques for software fault prediction". In: *Applied Soft Computing* 27 (2015), pp. 504–518.
4. Amod Kumar and Ashwni Bansal. "Software fault proneness prediction using genetic based machine learning techniques". In: *2019 4th International Conference on Internet of Things: Smart Innovation and Usages (IoT-SIU)*. IEEE. 2019, pp. 1–5.
5. Tim Menzies et al. "Implications of ceiling effects in defect predictors". In: *Proceedings of the 4th international workshop on Predictor models in software engineering*. 2008, pp. 47–54.
6. Tim Menzies et al. "Local vs. global models for effort estimation and defect prediction". In: *2011 26th IEEE/ACM International Conference on Automated Software Engineering (ASE 2011)*. IEEE. 2011, pp. 343–351.
7. Martin Shepperd et al. "Data quality: Some comments on the nasa software defect datasets". In: *IEEE Transactions on Software Engineering* 39.9 (2013), pp. 1208–1215.
8. J Sayyad Shirabad, Tim J Menzies, et al. "The PROMISE repository of software engineering databases". In: *School of information technology and engineering, University of Ottawa, Canada* 24 (2005), p. 3.
9. Eunsuk Chong, Chulwoo Han, and Frank C Park. "Deep learning networks for stock market analysis and prediction: Methodology, data representations, and case studies". In: *Expert Systems with Applications* 83 (2017), pp. 187–205.
10. Lei Qiao et al. "Deep learning based software defect prediction". In: *Neurocomputing* 385 (2020), pp. 100–110.
11. Khoa Phung, Emmanuel Ogunshile, and Mehmet Aydin. "A Novel Software Fault Prediction Approach To Predict Error-type Proneness in the Java Programs Using Stream X-Machine and Machine Learning". In: *2021 9th International Conference in Software Engineering Research and Innovation (CONISOFT)*. IEEE. 2021, pp. 168–179.
12. Khoa Phung, Emmanuel Ogunshile, and Mehmet Aydin. "Error-type–A Novel Set of Software Metrics for Software Fault Prediction". In: *IEEE Access* (2023).
13. Shyam R Chidamber and Chris F Kemerer. "A metrics suite for object oriented design". In: *IEEE Transactions on software engineering* 20.6 (1994), pp. 476–493.
14. Rudolf Ferenc et al. "An automatically created novel bug dataset and its validation in bug prediction". In: *Journal of Systems and Software* 169 (2020), p. 110691.
15. Paul Goodman. *The Practical Implementation of Software Metrics*. McGraw-Hill, Inc., 1993.
16. Jagdish Bansiya and Carl G. Davis. "A hierarchical model for object-oriented design quality assessment". In: *IEEE Transactions on software engineering* 28.1 (2002), pp. 4–17.
17. Rachel Harrison, Steve J Counsell, and Reuben V Nithi. "An evaluation of the MOOD set of object-oriented software metrics". In: *IEEE Transactions on Software Engineering* 24.6 (1998), pp. 491–496.
18. Thomas J McCabe. "A complexity measure". In: *IEEE Transactions on software Engineering* 4 (1976), pp. 308–320.
19. Maurice H Halstead. *Elements of Software Science (Operating and programming systems series)*. Elsevier Science Inc., 1977.
20. Dimitris Dranidis, Konstantinos Bratanis, and Florentin Ipate. "JSXM: A tool for automated test generation". In: *International Conference on Software Engineering and Formal Methods*. Springer. 2012, pp. 352–366.
21. P Kumudha and R Venkatesan. "Cost-sensitive radial basis function neural network classifier for software defect prediction". In: *The Scientific World Journal* 2016 (2016).
22. Xinli Yang et al. "Deep learning for just-in-time defect prediction". In: *2015 IEEE International Conference on Software Quality, Reliability and Security*. IEEE. 2015, pp. 17–26.

23. Yasutaka Kamei et al. "A large-scale empirical study of just-in-time quality assurance". In: *IEEE Transactions on Software Engineering* 39.6 (2012), pp. 757–773.
24. C Manjula and Lilly Florence. "Deep neural network based hybrid approach for software defect prediction using software metrics". In: *Cluster Computing* 22.Suppl 4 (2019), pp. 9847–9863.
25. Ting-Yan Yu, Chin-Yu Huang, and Neil C Fang. "Use of deep learning model with attention mechanism for software fault prediction". In: *2021 8th International Conference on Dependable Systems and Their Applications (DSA)*. IEEE. 2021, pp. 161–171.
26. Jean Petrić et al. "The jinx on the NASA software defect data sets". In: *Proceedings of the 20th International Conference on Evaluation and Assessment in Software Engineering*. 2016, pp. 1–5.
27. Tracy Hall et al. "A systematic literature review on fault prediction performance in software engineering". In: *IEEE Transactions on Software Engineering* 38.6 (2011), pp. 1276–1304.
28. Danijel Radjenović et al. "Software fault prediction metrics: A systematic literature review". In: *Information and software technology* 55.8 (2013), pp. 1397–1418.
29. David J Hand and Robert J Till. "A simple generalisation of the area under the ROC curve for multiple class classification problems". In: *Machine learning* 45 (2001), pp. 171–186.
30. Peter A Flach. "ROC analysis". In: *Encyclopedia of machine learning and data mining*. Springer, 2016, pp. 1–8.
31. Jason Osborne. "Improving your data transformations: Applying the Box-Cox transformation". In: *Practical Assessment, Research, and Evaluation* 15.1 (2010), p. 12.
32. Hidenori Ide and Takio Kurita. "Improvement of learning for CNN with ReLU activation by sparse regularization". In: *2017 international joint conference on neural networks (IJCNN)*. IEEE. 2017, pp. 2684–2691.
33. Nachiappan Nagappan et al. "Change bursts as defect predictors". In: *2010 IEEE 21st international symposium on software reliability engineering*. IEEE. 2010, pp. 309–318.

Blockchain Applications

A Step Towards Efficient Reduction of Ethereum's Turing-Completeness

Petar Trifunović, Dušan B. Gajić, Nebojša Horvat, Dinu Dragan, and Veljko B. Petrović

Abstract A blockchain's expressive power is ultimately equivalent to that of its underlying low-level language and virtual machine. The most popular blockchains are designed as either Turing-incomplete platforms, which intentionally restrict a blockchain's expressive power, or as Turing-complete systems with no theoretical restrictions to lower abstraction levels. Statistics show that 64% of applications on Ethereum—a Turing-complete blockchain—do not require a Turing-complete execution environment. On the other hand, none of the existing Turing-incomplete blockchains possess the expressive power to host those applications. In this paper, we propose an approach to creating a blockchain environment that represents the middle ground between the two extreme blockchain design approaches described. Instead of building a restrictive Turing-incomplete system from the ground up, as existing blockchains do, we take a reverse approach by reducing the Turing-completeness of Ethereum. Specifically, we introduce a more restrictive and secure way of creating loops in Ethereum's low-level language. To enable these restrictions, we alter both Ethereum's low-level language and its virtual machine architecture. As a result, these alterations make it highly unlikely, if not impossible, for infinite loops to occur and reduce the risk of unintentionally writing long-lasting finite loops. Furthermore, by choosing to reduce Ethereum, rather than building a new solution from scratch, we preserve the majority of Ethereum's high expressive power even after introducing restrictions. We show that the expressive power of our altered version of Ethereum is

P. Trifunović (✉) · D. B. Gajić · N. Horvat · D. Dragan · V. B. Petrović
University of Novi Sad, Faculty of Technical Sciences, Novi Sad, Serbia
e-mail: p.trifunovic@uns.ac.rs

D. B. Gajić
e-mail: dusan.gajic@uns.ac.rs

N. Horvat
e-mail: horva.n@uns.ac.rs

D. Dragan
e-mail: dinud@uns.ac.rs

V. B. Petrović
e-mail: pveljko@uns.ac.rs

N. Polatidis et al. (eds.), *Intelligent Distributed Computing XVII*, Studies in Computational Intelligence 1203, https://doi.org/10.1007/978-3-031-87639-4_11

at least equivalent to that of the set of primitive recursive functions, which is highly expressive despite its Turing-incompleteness.

Keywords Ethereum · Turing-completeness · Blockchain · Primitive recursive functions · Programming languages

1 Introduction

Blockchain is a distributed system controlled by all of its participants, instead of a privileged subset [46]. This concept of distributed control is known as decentralization [46]. The state of these distributed, decentralized systems is updated through immutable transactions, which, under the hood, commonly consist of a set of instructions from some low-level language. Choosing the low-level language that will govern the system's state transition during transactions is crucial for blockchain design. This language, and, consequently, the blockchain system it is used for, can either be Turing-complete, capable of implementing any algorithm, or Turing-incomplete,[1] meaning its expressive power and the set of algorithms it can implement is in some way restricted [20]. This paper proposes an approach to reducing the expressive power of a Turing-complete blockchain—Ethereum—by altering its underlying virtual machine (the Ethereum Virtual Machine, EVM) and its low-level language (the EVM bytecode), while preserving most of its ability to execute complex computational logic. We describe the first step in this reduction, which aims towards the prevention of infinite loops, since this is one potentially harmful functionality of Turing-complete languages [25]. To enable the reduction of Ethereum, we introduce certain changes to the EVM. There are two main aspects of these changes addressed in the paper:

- the alterations of the EVM bytecode instruction set and the extensions of the EVM architecture to support those alterations, and
- the way loops and branching constructs are implemented using the altered instruction set.

Our motivation for creating a Turing-incomplete blockchain is twofold. First, the statistics from [14, 28] show that at most 35.3% of verified Ethereum contracts require Turing-completeness. The amount of contracts that do not require Turing-completeness is almost twice as big. These statistics [14, 28] make it reasonable to explore ways in which Turing-completeness of EVM can be reduced, while aiming to preserve most of its expressive power. Second, Turing-completeness increases the risk of developers making errors and creating security flaws. One of the biggest Ethereum hacks, the DAO hack [7], was a consequence of the oversights developers made while writing contracts, but was enabled by the Turing-completeness of the

[1] The terms *Turing-incomplete* and *non-Turing complete* can be used as synonyms.

underlying low-level EVM. Reducing the Turing-completeness of the EVM bytecode could lower the occurrence of such mistakes during Ethereum smart contract development.

Infinite loops and impractically long finite loops are prevented in the EVM by its code-execution billing system, where each EVM bytecode instruction consumes a certain amount of gas [10]. Contracts which consume too much eventually run out of available gas and are aborted. The EVM only prevents the execution of such too-long-lasting loops, but does not prevent developers from writing them. The sole possibility of writing such loops is undesirable and can lead to creating smart contracts which abort unexpectedly. Our proposed alterations of the EVM make it impossible to write loops the usually way in the EVM bytecode, using jumping instructions. Consequently, these alterations make it very hard, if not impossible, to create infinite loops. Also, writing finite loops in our modified version of the EVM bytecode requires explicit setting of the number of iterations, reducing the risk of writing loops with impractically large amount of iterations.

Our approach to creating a Turing-incomplete blockchain differs from the existing solutions. The existing solutions are mostly based on the first blockchain implementation, Bitcoin [21]. Bitcoin uses a strict, Turing-incomplete, low-level language, called Script [4]. This language, and those used by other similar blockchains, is built from the ground-up, by including the functionalities necessary to support a specialized transaction set. However, there are no documented examples of blockchain platforms being made Turing-incomplete by reducing a Turing-complete one, rather than building a new one from scratch. This reduction-based approach is exactly what we propose. Our solution starts from the Turing-complete Ethereum blockchain, and reduces certain aspects of it, while aiming to preserve most of its expressive power. As the first step towards turning Ethereum into an efficient Turing-incomplete system, we aim towards the prevention of infinite loops in the EVM.

Our method of preventing infinite loops in the EVM bytecode is based on a conceptual language proposed in [19] which can be used to implement so-called Loop programs. It consists of a small instruction set focused on implementing loops with the number of iterations known in advance. It implements the set of primitive recursive functions. The definition of these functions, as well as the proof of equivalence between the set of functions computable by Loop programs and the primitive recursive set are both given in [19]. Although the set of primitive recursive functions is not Turing-complete, a large amount of algorithms can be implemented using them. In fact, as described in [15, 19], the expressiveness of primitive recursion is very high and it is not trivial to devise an algorithm which cannot be converted to a primitive-recursive form. The expressiveness of primitive recursion and its equivalence to Loop programs indicates that the changes we introduce to the EVM instruction set based on the instruction set described in [19] leaves EVM with the power of implementing complex computational logic.

The rest of the paper is structured as follows. First, Sect. 2 gives an overview of the related work. Then, Sects. 3 and 4 focus on the technical aspects of Loop programs and the EVM, respectively. Section 5 describes our proposed alteration of the EVM instruction set and its architecture. Section 6 presents how important language constructs can be implemented using the altered EVM architecture and instruction set. Section 7 gives final conclusions and describes potential points of further research.

2 Related Work

The first implementation of a blockchain platform is the Bitcoin, proposed in 2008 [21]. It is used as a decentralized digital asset management network. The state of the network changes through transactions, whose logic is expressed in a Turing-incomplete language, called Script [4]. This language governs the way in which Bitcoin assets are transferred. Bitcoin uses the concept of Unspent Transaction Output (UTXO), which represents a unit of digital asset that remains in the network after a transaction and can be used as input to another one. Since Bitcoin is not meant to enable execution of complex logic, it does not require Turing-completeness.

The birth of Ethereum [44] enabled blockchain networks to host distributed applications [6], implemented in the form of smart contracts [3]. This requires an underlying state machine capable of executing logic much more complex than that of Bitcoin's transactions. Therefore, Ethereum was created as a virtual state machine, called the Ethereum Virtual Machine (EVM). The EVM's low-level language consists of a set of operational instructions—*opcodes*. Each of these opcodes is one byte long, which is why they are often referred to as *bytecodes*.[2] The EVM and its bytecode language are Turing-complete, enabling the creation of more constructs unsupported in Bitcoin, such as recursion and infinite loops.

In this section, we give an overview of some of the most popular blockchains, in terms of their capabilities and the Turing-completeness of their low-level layers.

A survey of the various blockchain platforms and their properties, including Turing-completeness, is given in [35]. In [35], Stellar is listed as Turing-incomplete, since this survey was done in 2021. However, in 2022 Stellar added support for Turing-complete smart contracts. Stellar [33] is one of the first blockchain platforms, released in 2014, and it targeted the problems Bitcoin had with scalability and transaction settlement times. At first, it supported Turing-incomplete smart contracts, but Stellar was faster than Ethereum in transferring digital assets, which was indeed its main purpose [45]. Later, Stellar switched to Soroban [34]—a Turing-complete platform for developing Rust-based smart contracts.

The Algorand blockchain [1] had a path similar to Stellar. At first, it supported only simple on-chain logic. Later, support for a higher-level smart contract programming language named TEAL was announced [22], but it was Turing-incomplete and focused on developing specialized smart contracts for decentralized finances. Eventually, the TEAL language and the Algorand Virtual Machine (AVM) were extended to support Turing-complete logic [41].

[2] The terms *opcode* and *bytecode* are not specific to Ethereum or blockchain, but are described here for convenience.

Tezos [36] is a Turing-complete blockchain that uses a low-level language called Michelson [23]. Like Ethereum, Tezos uses a stack-based virtual machine, but, unlike EVM bytecode, Michelson uses high-level data types and strict static type-checking. These properties of the Michelson language make Tezos more suitable for formal verification and execution analysis [24].

Solana [31] is a blockchain which uses a compiler and a low-level language which were not initially intended for blockchain purposes. It uses the LLVM compiler infrastructure [17] to provide intermediate-level code optimizations, and a modified, Turing-complete implementation of the Extended Berkeley Packet Filter (eBPF) bytecode [32]. Solana's high-level smart contracts are usually written in Rust, and, in order to support Rust's expressiveness, the contracts must translate to a Turing-complete low-level language, such as ePBF.

Many blockchain platforms use Ethereum's virtual machine, and build the rest of the platform on top of it. On a source-code level, some of these use general-purpose languages, such as Python or Rust, while others use languages specialized for smart contracts, like Solidity [38] or Vyper [40]. While the high-level language may differ, they all compile into Ethereum's Turing-complete bytecode. Examples of such blockchains are Avalanche [2], Fantom [12], Tron [39], Polygon [26].

Blockchains that do not have native support for decentralized applications are most commonly used as platforms for digital assets, and often use some form of Bitcoin's UTXO concept. These platforms cannot be considered Turing-complete machines. Apart from Bitcoin, popular examples are Litecoin [16] and Dogecoin [8], both of which are open-source digital currencies, and they do not inherently support the development of decentralized applications.

In conclusion, two radically different approaches to blockchain design exist. One intentionally limits the capabilities of a blockchain platform, decreasing its Turing-completeness, making such blockchains less prone to bugs, errors, and security attacks, and easier to specialize for specific use cases [25, 28], but also incapable of executing complex application logic [13]. The other approach uses Turing-complete environments to enable the execution of powerful programs and complex application logic [13], which, on the other hand, poses higher security risks and increases the chances of bugs and errors [25, 28]. As stated in [14, 28], at least 64% of smart contracts deployed and verified on the Ethereum blockchain do not require Turing-completeness, which makes efforts to devise a Turing-incomplete environment for their execution worthwhile.

Our proposal to reduce the EVM's strength step-by-step, rather than designing a Turing-incomplete environment from the ground up, may produce a starting point for a solution that stands in the middle-ground between the described extremes. The description of this reduction process given in the following sections and the overview of existing blockchain platforms given in this section should provide an insight into the position of our approach in the existing blockchain ecosystem.

3 Loop Language and Primitive Recursion

In [19], Meyer and Ritchie described a conceptual programming language consisting of a minimalist set of instructions capable of implementing the set of primitive recursive functions [29]. No name was given to this programming language in [19], but the programs implemented in it were called *Loop programs*. For convenience, in the rest of this paper, we refer to the language as the *Loop language*. Loop is a register-based language, with no limits to the number of registers and the size of integers they can hold. The language consists of five types of instructions, but the most relevant to this paper are the $LOOP X$ and END types, where *LOOP* and *END* are the language's keywords, and *X* represents a register. Each *LOOP* must have a matching *END* instruction. If the value in register *X* is an integer x, the block of code between *LOOP* and *END* is executed x times. These instructions do not change the value in *X* themselves, and explicitly changing the value of *X* within the repeating block of code does not interfere with the number of iterations to be executed.

The formal description of primitive recursive functions consists of a set of initial functions and rules by which those functions can be used to create more complex ones [19, 29]. In a less formal manner, it is the set of functions computable by algorithms in which every loop is a *for*-like loop with the number of iterations known in advance. A good rationale for this is given on Wikipedia's page for primitive recursion [43]—primitive recursion can be interpreted as a set of rules describing how one iteration of a function depends on the previous iteration, and how the function behaves in the first iteration. These rules are sufficient to implement every primitive recursive function of $n + 1$ variables, $f(x_1, ..., x_n, k)$, as a for-loop of k iterations [43].

Meyer and Ritchie [18, 19] proved that the set of functions computable by Loop programs is actually the set of functions defined by primitive recursion. The set of primitive recursive functions cannot express all computable functions, meaning it is not Turing-complete. Nevertheless, a very large portion of functions can be represented in a primitive recursive form [15], indicating that Loop programs are capable of implementing complex algorithms.

4 Ethereum Virtual Machine

Ethereum was designed as a distributed, stack-based, state machine, enabling application development in a blockchain environment, rather than merely representing a cryptocurrency payment system [27]. The Ethereum Virtual Machine (EVM) is the engine that defines the rules according to which Ethereum changes its states [9]. These rules are formally defined in the Ethereum Yellowpaper [44]. There are multiple implementations of the EVM, all of which are compatible as long as they adhere to the Yellowpaper specification [11].

EVM state changes are triggered by executing bytecode programs consisting of a special set of instructions, symbolically represented by EVM opcodes. The next instruction to execute is determined by the contents of the so-called *program counter* (PC). The counter increments by 1 after most of the instructions. Instructions that differ from the rest in how they change the PC are:

- *JUMP*, which unconditionally changes the PC to the value on top of the stack,
- *JUMPI*, which checks the second value on the stack, and, if it is not equal to 0, changes the PC to the value on top of the stack; otherwise, the PC increments by 1, and
- *PUSHx*, where the x can be any number in the range from 0 to 32; *PUSHx* increments the PC by $1 + x$; each *PUSHx* instruction pushes an x bytes long value to the stack.

The EVM is a Turing-complete machine, but it solves the halting problem [5] by introducing the concept of gas [10]. The execution of every EVM instruction consumes a certain amount of gas. Thus, an infinite loop would require an infinite amount of gas. Whoever invokes an EVM program must provide enough gas for its execution, and the impossibility of anyone possessing an infinite amount of gas prevents infinite loops, or any other impractically long algorithm, from being executed in an EVM blockchain environment [10].

5 Extending the EVM with the Loop Language Instruction Set

Preventing infinite loops in the EVM's low-level language inevitably introduces changes to the the EVM's abstract machine architecture and its instruction set. In the following subsections, we analyze which EVM bytecode instructions should be removed to prevent infinite loops and propose a way in which they can be replaced with instructions similar to those from the Loop language. Consequently, we introduce certain extensions to the EVM architecture to support proper execution of the new instruction set.

5.1 Detecting the EVM Instructions to Remove

To determine which instructions from the EVM's instruction set should be removed or replaced by instructions from the Loop language, it was necessary to create an EVM program which contains an infinite loop. To easily generate such a program, we wrote a simple smart contract using the high-level programming language *Solidity* [38] containing only one function with an infinite, body-less *while* loop—$while(true)\{\}$. We compiled this code to EVM bytecode using version 0.8.24 of the *solc* compiler [37]. This produced an EVM bytecode where the infinite loop is implemented using only six EVM instructions, as shown in Fig. 1. The numbers prefixed with colon signs represent the hexadecimal addresses of the instructions. The *PUSH* instructions span a

```
...
JUMPDEST      :0x33
PUSH1 0x1     :0x34–0x35
PUSH1 0x33    :0x36–0x37
JUMPI         :0x38
...
```

Fig. 1 EVM opcodes of an infinite while loop

range of addresses, since their immediate operands occupy the number of bytes equal to the suffix of the *PUSH* instruction. The instructions in the code do the following:

1. *JUMPDEST* only marks a valid destination to jump to in the program.
2. *PUSH1 0x1* pushes number 1 to the stack. This value represents the *true* value from the condition part of the *while* loop.
3. *PUSH1 0x33* pushes hexadecimal value 33 to the stack, which represents the address of the previously mentioned *JUMPDEST* instruction.
4. *JUMPI* instruction checks the second value on the stack and, if it is not zero, jumps to the address on top of the stack. In our example, this second value on the stack will always be equal to 1, and the program will always jump to address 0x33 and perform the same set of instruction all over again.

Further details about these EVM instructions are available in [44]. The instruction flow in this EVM program indicates that the main building block of infinite loops are jump instructions. In this case, the jump instruction used is a conditional one, but it essentially implements an unconditional jump that occurs every time this instruction is executed. Thus, our approach to preventing such loops is to eliminate the jump instructions, both unconditional and conditional ones. The EVM has only two jump instructions—*JUMPI*, already explained above, and *JUMP* which performs an unconditional jump to the address stored at the top of the stack.

5.2 Extending the EVM Architecture and Instruction Set

The alteration of the EVM instruction set proposed in this paper means the complete removal of *JUMP* and *JUMPI* instructions, and the introduction of *LOOP-END* instruction pair similar to that presented in [19]. The problem with jump instructions is that they enable arbitrary updates of the program counter. Despite being named *JUMP* and *JUMPI* to better correspond to their abstract behavior, what these instructions actually do on an architectural level is directly alter the value stored in the program counter. Eliminating these instructions from the EVM instruction set leaves us with instructions whose impact on the program counter is hidden behind the definition and the state transition mechanism of the virtual machine, instead of being directly exposed through the EVM language syntax. Such implicit behavior is desirable in our case, as the exposure of the program counter through jump instructions is what makes infinite loops possible to implement. Thus, it is necessary to ensure the security of the program counter after introducing *LOOP* and *END* to the EVM instruction set.

The *LOOP* and *END* instructions proposed in [19] need to be modified to some extent to fit the EVM architecture. However, for these instructions to work properly, we also propose certain extensions of the EVM architecture itself.

The *LOOP* and *END* instructions need to keep track of how many times the block embedded between them has been executed. In the original Loop language, this value is passed to the *LOOP* instruction via a register, and changing the value of that register during the loop's execution does not change the number of iterations to be executed. Since EVM is stack-based, registers do not exist in its architecture, so an alternative must be found. This alternative must guarantee that the value passed to the *LOOP* instruction cannot be altered in any way that interferes with the correct number of iterations. None of the data locations in the EVM architecture [30, 44] are capable of providing this guarantee. Stack, storage, and memory are all exposed to the rest of the EVM program through other bytecode instructions, and, if any of these data locations was chosen, there would be no way to prevent the instructions from inside the loop's body from tampering with the correct number of iterations left:

- **Stack** can be accessed using a large number of EVM instructions, since most of them read inputs from the stack, or write their result on it.
- **Storage** is expensive in terms of gas its usage requires, and the values it holds can be accessed using *SLOAD* and changed using the *SSTORE* instruction.
- **Memory** is cheaper than storage, but it is more expensive than stack, and can be read from using *MLOAD*, and written to using the *MSTORE* and *MSTORE8* instructions.

Unlike these three data locations, the *calldata* location cannot be altered using other EVM instructions, but it cannot be changed at all, since it is read-only. Therefore, if the number of iterations were stored in the *calldata* location, it would be impossible to properly update it.

Due to all the described inadequacies of the existing data locations, we propose a stack-like structure, kept, in an abstract sense, in the same part of the EVM architecture as the program counter. Sharing the location with the program counter would make this stack as protected from the EVM instructions as the program counter. It would not be directly accessible through the EVM instruction set, but would only be used implicitly by the *LOOP* and *END* instructions. This usage, as well as the way these instructions update the rest of the EVM state, would be managed by the EVM itself. Before describing how *LOOP* and *END* update the machine state and their overall logic, we introduce the following symbols:

- ***S***, referring to the standard stack allocated to every EVM program,
- ***S****, referring to the newly-introduced, protected stack, used to keep track of loops and their execution,
- ***PC***, referring to the EVM program counter.

If all the loops in a program had at least one iteration, the logic behind *LOOP* and *END* would be straightforward—a *LOOP* instruction would push the number of iterations to $S*$, and, on top of it, the address of the next instruction in the program. The *END* instruction would check the second value in $S*$ (the number of iterations of the current loop), and, if greater than zero, would jump to the address equal to the top value in $S*$ (the address of the first instruction after the latest encountered *LOOP*). However, zero-iteration loops could not be properly resolved, since the logic behind the described mechanism is equivalent to the logic of a *do-while* loop, where at least one iteration is executed before the exit condition is checked for the first time. Because of this problem, we add a 32-byte immediate operand to the *LOOP* instruction, representing the address of the *END* instruction paired with it. This makes it possible to jump over the body of the loop straight to the *END* instruction, enabling proper execution of zero-iteration loops. In order to prevent a backward jump, and, in turn, infinite loops, this address must be greater than the address of the corresponding *LOOP* instruction, that is, greater than the value of PC when *LOOP* is encountered. Also, it must be the address of an *END* instruction, similar to how the original *JUMP* and *JUMPI* must jump to a *JUMPDEST*. These conditions must be met, otherwise, the program would generate an error during runtime. Since the original jump instructions read the destination as a value from the stack, and each slot in the stack is 32 bytes long, the length of this operand is also equal to 32 bytes. We refer to the value of this immediate operand as *ENDAddr*.

With these extensions to the EVM architecture and instruction set, the behaviour of *LOOP* and *END* can be described in more detail. When a *LOOP* instruction is encountered:

- the top value (the number of iterations) in S is popped out of S and pushed to $S*$,
- a value equal to $PC + 33$ is pushed to $S*$; this value is equal to the address of the first instruction after *LOOP*; $33 = 1 + 32$ is added to PC because this is how far away the next instruction is from *LOOP*—1 for the length of the *LOOP* instruction itself, and 32 for the length of *ENDAddr*, and
- PC is set to *ENDAddr*.

As explained above, *LOOP* reads the number of iterations from S. It is up to the programmer or a higher-level compiler to place this value on the stack in advance.

When an *END* instruction is encountered:

- if the second value in $S*$ (the number of iterations left) is greater than zero, it is reduced by 1 and PC is set to the top value in $S*$;
- if the second value in $S*$ is equal to or less than zero, the top two values are removed from $S*$, and the PC is incremented by 1.

The addition of an immediate operand to the *LOOP* instruction makes room for structuring the code in ways not possible in the original Loop language. This is due to the fact that a *LOOP* instruction is paired with *END* by the *ENDAddr* operand, rather than by the syntax itself. This means that two or more *LOOP* instructions can share an *END* instruction. This can even be convenient in cases in which there are

```
...
LOOP ENDAddr1
          ...
          LOOP ENDAddr2
          ...
END                :ENDAddr1
...
          END      :ENDAddr2
...
```

Fig. 2 Interleaved loops

no other instructions between a number of consecutive *END* instructions belonging to a group of nested loops. In this case, the consecutive *END*s can be reduced to one. Pairing *LOOP-END* by address enables loop interleaving. Although this can make the code appear logically complex, it does not interfere with proper program execution. This is because the only point of connection between a *LOOP* and an *END* is during loop initialization. Which *END* instruction reduces the number of iterations of a loop makes no difference, as long as that number is eventually reduced to zero. Figure 2 shows an example of interleaved loops. Although the second *LOOP* jumps over *ENDAddr1* to reach its *END*, it cannot jump to a position with no *END* instruction. Despite being interleaved, there is no way for the execution to go past *ENDAddr2* (except to another *END* instruction further down the code) before both loops have no iterations left.

6 Alternative Implementation of Language Constructs

After removing jumps from the EVM instruction set, two language constructs that use jumps in their original implementation require alternative implementations—loops and branches. We demonstrate these alternatives using representative EVM code examples. To generate these EVM programs, we compiled *Solidity* smart contracts[3] to the EVM bytecode using version 0.8.24 of the *solc* compiler [37]. For each example, we compare the original, *solc*-generated EVM code with the code modified according to our EVM bytecode and architecture changes.

We start with a simple example—an empty *for* loop with three iterations. Figure 3a shows the original EVM code of this loop. First, 0 is pushed to the stack as the initial iteration counter value, followed by a *JUMPDEST*. Then, 3 is pushed to the stack, as the iteration number limit. The next five instructions check if the bound is reached and, if so, jump to address 0x48, located later in the code. Otherwise, the rest of the instructions before *PUSH1 0x34* increase the iteration counter and clean the stack.

[3] We do not show the *Solidity* smart contracts because their content is not of great importance for this paper. *Solidity* was used only because it is more convenient to compile it to the EVM bytecode, than to write the bytecode directly.

```
...
PUSH0
JUMPDEST
PUSH1 0x3
DUP2
SLT
ISZERO
PUSH1 0x48
JUMPI
DUP1
DUP1
PUSH1 0x1
ADD
SWAP2
POP
POP
PUSH1 0x34
JUMP
...
```

(a) Original code

```
...
PUSH0
PUSH1 0x3
LOOP ENDAddr
    DUP1
    DUP1
    PUSH1 0x1
    ADD
    SWAP2
    POP
    POP
END      :ENDAddr
...
```

(b) Altered code

Fig. 3 Empty for loop

The execution then jumps to the *JUMPDEST* at the beginning of Fig. 3a. In the fourth iteration, the counter will be equal to 3 and the execution will jump to 0x48. Figure 3b shows the altered code. The *PUSH1 0x3* instruction sets the number of iterations for the *LOOP* instruction. The *LOOP* and *END* instructions keep track of the number of iterations. The code inside the *LOOP-END* block increments a counter, but this does not affect the iteration tracking. If the counter has no meaning outside of the loop, the *LOOP-END* block can be left empty. Note that *ENDAddr* symbol is used both here and in the following examples instead of a hexadecimal address, for convenience.

The next example is a slightly more realistic *for* loop. It iterates through an array and is limited by the array's length. In each iteration, an element of the array is assigned a constant value. For convenience, in Fig. 4 we only show the part of code that checks if the loop is finished and the part that updates the iterator. The first part in Fig. 4a sets up the iterator value, inserts a *JUMPDEST*, and loads the array's length using *DUP2* and *MLOAD*. The next five instructions check if the iterator value is lower than the array length, and, if not, jump to address 0x3D, which is the address of the last *JUMPDEST* in Fig. 4a. This jump ends the loop. The rest of the code increases the iterator variable by one and jumps to address 0x108, which is the first *JUMPDEST* in the figure. Most of the code in Fig. 4a is replaced by *LOOP* and *END* in Fig. 4b. The first *JUMPDEST* is eliminated, since jump instructions have no purpose in our altered EVM. After *MLOAD* loads the length of the array to the stack, the *LOOP* instruction can be inserted, reading the number of iterations from the stack. All the instructions between *MLOAD* and the middle ellipsis in Fig. 4a are

```
...
PUSH0
JUMPDEST
DUP2
MLOAD
DUP2
LT
ISZERO
PUSH2 0x13D
JUMPI
...
DUP1
PUSH1 0x1
ADD
SWAP1
POP
PUSH2 0x108
JUMP
JUMPDEST
...
```

(a) Original code

```
...
PUSH0
DUP2
MLOAD
LOOP ENDAddr
  ...
END     :ENDAddr
...
```

(b) Altered code

Fig. 4 A *for* loop iterating through an array

replaced by *LOOP*. Similarly, the second part of the original code, which updates the iterator variable, is replaced by the *END* instruction.

Translating a loop to the modified EVM is straightforward. However, branching, originally handled by jump instructions, must now be implemented using loops. As explained on the Wikipedia page of the Loop programming language [42], branching is implemented with two loops. Based on the branching condition, one loop gets one iteration, while the other gets none. The "branch" that should be executed is the loop that executes one time. For *if-then* branching, a single loop which executes one, or zero times, depending on the condition, is sufficient.

Figure 5 shows a conditional branch in the EVM bytecode, in its original and altered form. The execution of this code stops if the top stack value at the beginning of the presented code is non-zero.The original code in Fig. 5a duplicates the top of the stack using *DUP1*, and compares it to zero using *ISZERO*. If it is zero, *JUMPI* jumps to the *JUMPDEST* at the end of the code. Otherwise, the next three instructions halt the program. Figure 5b shows how this code is implemented with the new instruction set. The loop contains the three instructions from the original code which stop the execution. This loop should have either 0 or 1 iteration, depending on the value on top of the stack at the beginning of the code. If the top of the stack is 0, *LOOP* should have 0 iterations, and if not, it should have 1. However, *ISZERO* does the opposite—places 1 on the stack if the current top of the stack is equal to 0. This is why another *ISZERO* is used to invert the result of the first one. After this, *LOOP*

```
...
DUP1
ISZERO
PUSH1 0xE
JUMPI
PUSH0
DUP1
REVERT
JUMPDEST
...
```

(a) Original code

```
...
DUP1
ISZERO
ISZERO
LOOP ENDAddr
    PUSH0
    DUP1
    REVERT
END      :ENDAddr
...
```

(b) Altered code

Fig. 5 Conditional branching example

can read the correct number of iterations, enabling proper simulation of conditional branching.

7 Conclusions

In this paper, we propose an approach to reducing the Turing-completeness of the EVM bytecode, while preserving most of its expressive power. Our motivation stems from statistics showing that almost two thirds of verified Ethereum smart contracts do not require Turing-completeness. We demonstrate how the EVM can be altered to replace jumping instructions with certain instructions from the Loop language. This modification reduces the risk of writing infinite loops and finite loops with too many iterations. These restrictions on loop creation could turn Ethereum into an environment less prone to unexpected development errors, without a large negative impact on Ethereum's high level of expressiveness.

Our work leaves room for further research on efficient Turing-incomplete blockchains. Our proposed changes to the EVM are only one step towards creating an efficiently reduced, yet powerful version of this virtual machine. Further analysis of the structure of Ethereum smart contracts is needed to determine the subset of EVM's capabilities which these contracts actually require. Potential benefits of our solution include improved security, better static code analysis, and easier gas consumption estimation. All of these benefits could come in the future, and they generate great space for more work on this topic.

Disclosure of Interests The authors have no competing interests to declare that are relevant to the content of this article.

References

1. Algorand: Algorand Technologies, http://algorandtechnologies.com/, Accessed on 14.05.2024
2. Avalanche Foundation: Avalanche: Create Without Limits – dApp Platform, http://www.avax.network/, Accessed on 14.05.2024
3. Bartoletti, M., Pompianu, L.: An Empirical Analysis of Smart Contracts: Platforms, Applications, and Design Patterns. In: Financial Cryptography and Data Security. pp. 494–509. Springer International Publishing (2017). https://doi.org/10.1007/978-3-319-70278-0_31
4. Bitcoin Wiki: Script - Bitcoin Wiki, http://en.bitcoin.it/wiki/Script, Accessed on 14.05.2024
5. Burkholder, L.: The Halting Problem. ACM SIGACT News **18**(3), 48–60 (1987https://doi.org/10.1145/24658.24665
6. Buterin, V.: Ethereum White Paper: A Next Generation Smart Contract & Decentralized Application Platform (2014), https://ethereum.org/en/whitepaper/, Accessed on 10.09.2024
7. Cryptopedia Staff: The DAO: What Was the DAO Hack? – Gemini, http://www.gemini.com/cryptopedia/the-dao-hack-makerdao, Accessed on 14.05.2024
8. Dogecoin: Dogecoin - An open-source peer-to-peer digital currency, http://dogecoin.com/, Accessed on 14.05.2024
9. Ethereum Foundation: Ethereum Virtual Machine (EVM) – ethereum.org, http://ethereum.org/developers/docs/evm, Accessed on 14.05.2024
10. Ethereum Foundation: Gas and Fees – ethereum.org, http://ethereum.org/en/developers/docs/gas/, Accessed on 14.05.2024
11. Ethereum Foundation: Nodes and clients – ethereum.org, http://ethereum.org/developers/docs/nodes-and-clients#execution-clients, Accessed on 14.05.2024
12. Fantom Foundation: Fantom, http://fantom.foundation/, Accessed on 14.05.2024
13. Hu, B., Zhang, Z., Liu, J., Liu, Y., Yin, J., Lu, R., Lin, X.: A Comprehensive Survey on Smart Contract Construction and Execution: Paradigms, Tools, and Systems. Patterns **2**(2), 100179 (2021).https://doi.org/10.1016/j.patter.2020.100179
14. Jansen, M., Hdhili, F., Gouiaa, R., Qasem, Z.: Do Smart Contract Languages Need to Be Turing Complete? In: Prieto, J., Das, A.K., Ferretti, S., Pinto, A., Corchado, J.M. (eds.) Blockchain and Applications. pp. 19–26. Springer International Publishing, Cham (2020)
15. Kahrs, S.: The Primitive Recursive Functions are Recursively Enumerable, https://www.cs.kent.ac.uk/people/staff/smk/primrec.pdf, Accessed on 10.09.2024
16. Litecoin Project: Litecoin - Open source P2P digital currency, http://litecoin.org/, Accessed on 14.05.2024
17. LLVM Foundation: The LLVM Compiler Infrastructure Project, http://llvm.org/, Accessed on 14.05.2024
18. Meyer, A.R., Ritchie, D.M.: Computational Complexity and Program Structure. IBM Watson Research Center Yorktown Heights, New York (1967)
19. Meyer, A.R., Ritchie, D.M.: The Complexity of Loop Programs. In: Proceedings of the 1967 22nd National Conference. p. 465-469. ACM '67, Association for Computing Machinery, New York, NY, USA (1967).https://doi.org/10.1145/800196.806014
20. Michaelson, G.: Programming Paradigms, Turing Completeness and Computational Thinking. The Art, Science, and Engineering of Programming **4**(3) (Feb 2020https://doi.org/10.22152/programming-journal.org/2020/4/4
21. Nakamoto, S.: Bitcoin whitepaper (2008), http://bitcoin.org/bitcoin.pdf
22. Nelson, D.: Algorand 2.0's New Non-Turing-Complete Smart Contracts Are a Feature, Not a Bug - CoinDesk, http://www.coindesk.com/tech/2019/11/22/algorand-20s-new-non-turing-complete-smart-contracts-are-a-feature-not-a-bug/, Accessed on 14.05.2024
23. Nomadic Labs: Michelson: the language of Smart Contracts in Tezos - Tezos (master branch, 2024/03/29 09:09) documentation, http://tezos.gitlab.io/active/michelson.html, Accessed on 14.05.2024
24. OpenTezos: Introduction – OpenTezos, http://opentezos.com/formal-verification, Accessed on 14.05.2024

25. Pickard, G.: Programming languages shouldn't and needn't be Turing complete – Gabriel Pickard, http://www.gabrielpickard.com/posts/programming-languages-shouldnt-and-neednt-be-turing-complete/, Accessed on 14.05.2024
26. Polygon Labs UI (Cayman) Ltd: Web3, Aggregated., http://polygon.technology/, Accessed on 14.05.2024
27. Rankhambe, B.P., Kaur Khanuja, H.: A Comparative Analysis of Blockchain Platforms - Bitcoin and Ethereum. In: 2019 5th International Conference On Computing, Communication, Control And Automation (ICCUBEA). pp. 1–7 (2019https://doi.org/10.1109/ICCUBEA47591.2019.9129332
28. Rautenberg, M.J., Rezabek, F.: A Case Study of Security Vulnerabilities in Smart Contracts. In: Proceedings of the Seminar Innovative Internet Technologies and Mobile Communications (IITM) (2022).https://doi.org/10.2313/NET-2022-11-1_10
29. Robinson, R.M.: Primitive Recursive Functions. Bulletin of the American Mathematical Society **53**(10), 925 – 942 (1947)
30. smlXL, Inc: EVM Codes - About the EVM, http://www.evm.codes/about#executionenv, Accessed on 14.05.2024
31. Solana Foundation: Web3 Infrastructure for Everyone – Solana, http://solana.com/, Accessed on 14.05.2024
32. Solana Labs: solana-labs/rbpf: Rust virtual machine and JIT compiler for eBPF programs, http://github.com/solana-labs/rbpf/, Accessed on 14.05.2024
33. Stellar Development Foundation: Stellar – A Blockchain Network for Payments and Tokenization, http://stellar.org/, Accessed on 14.05.2024
34. Stellar Development Foundation: Stellar – Soroban – Smart Contracts Platform on Stellar, http://stellar.org/soroban, Accessed on 14.05.2024
35. Suvitha, M., Subha, R.: A Survey on Smart Contract Platforms and Features. In: 2021 7th International Conference on Advanced Computing and Communication Systems (ICACCS). vol. 1, pp. 1536–1539 (2021)https://doi.org/10.1109/ICACCS51430.2021.9441970
36. Tezos Foundation: Home – Tezos, http://tezos.com/, Accessed on 14.05.2024
37. The Solidity Authors: Installing the Solidity Compiler - Solidity 0.8.24 documentation, http://docs.soliditylang.org/en/latest/installing-solidity.html, Accessed on 14.05.2024
38. The Solidity Authors: Solidity - Solidity 0.8.24 documentation, https://docs.soliditylang.org/en/v0.8.24/, Accessed on 14.05.2024
39. TRON Tech Limited: TRON – Decentralize The Web, http://tron.network/, Accessed on 14.05.2024
40. Vyper Team: Vyper - Vyper documentation, http://docs.vyperlang.org, Accessed on 14.05.2024
41. Weathersby, J.: Introducing Algorand Virtual Machine: AVM 0.9 Release – Algorand Developer Portal, http://developer.algorand.org/articles/introducing-algorand-virtual-machine-avm-09-release/, Accessed on 14.05.2024
42. Wikipedia: LOOP (programming language) - Wikipedia, http://en.wikipedia.org/wiki/LOOP_(programming_language), Accessed on 14.05.2024
43. Wikipedia: Primitive recursive function - Wikipedia, http://en.wikipedia.org/wiki/Primitive_recursive_function, Accessed on 14.05.2024
44. Wood, G., et al.: Ethereum: A secure decentralised generalised transaction ledger (2014), https://ethereum.github.io/yellowpaper/paper.pdf
45. XLM Wallet: Stellar vs Ethereum smart contracts: what's the difference? – by XLM Wallet – Medium, http://medium.com/@XLMwalletCo/stellar-vs-ethereum-smart-contracts-whats-the-difference-4dee5a7c044d, Accessed on 14.05.2024
46. Zarrin, J., Wen Phang, H., Babu Saheer, L., Zarrin, B.: Blockchain for Decentralization of Internet: Prospects, Trends, and Challenges. Cluster Computing **24**(4), 2841–2866 (Dec 2021).https://doi.org/10.1007/s10586-021-03301-8

Visualizing Bitcoin's Evolution: Dimensionality Reduction and Stylized Facts Techniques

Huma Zafar and Stelios Kapetanakis

Abstract This study examines Bitcoin's market behavior from May 2018 to June 2023, employing advanced statistical techniques to uncover its underlying characteristics. The analysis reveals several key findings: Bitcoin's returns exhibit a non-normal distribution, indicating a higher likelihood of extreme price movements. Periods of high volatility tend to persist in Bitcoin's market. Past price movements have limited predictive power for future returns. Bitcoin's volatility is not constant and negative price shocks often lead to increased volatility. Outlier analysis identifies significant market events impacting Bitcoin's price. These findings underscore the volatile and unpredictable nature of Bitcoin, emphasizing the need for robust risk management strategies. Future research should explore comparative dynamics with other cryptocurrencies, employ time-series and machine learning techniques, and assess the impact of macroeconomic factors and regulatory changes to enhance understanding and investment strategies in the cryptocurrency market.

Keywords Neural networks · Blockchain · Deep learning

1 Introductions

Bitcoin, introduced in 2009, has grown from a niche digital currency into a significant asset class within global financial markets. Its decentralized nature provides an alternative to traditional fiat currencies, attracting both retail and institutional investors. However, Bitcoin's volatility, marked by frequent and unpredictable price swings, presents challenges for market participants and researchers who aim to understand its driving factors.

H. Zafar (✉) · S. Kapetanakis
Middlesex University, London, England
e-mail: huma@dstr.co.uk

Distributed Analytics Solutions, London, UK

S. Kapetanakis
e-mail: stelios@dstr.co.uk

N. Polatidis et al. (eds.), *Intelligent Distributed Computing XVII*, Studies in Computational Intelligence 1203, https://doi.org/10.1007/978-3-031-87639-4_12

Therefore, this study seeks to examine Bitcoin's market behavior by analyzing a dataset from May 2018 to June 2023. Employing advanced statistical techniques, including PCA and t-SNE, the study investigates key price trends and market anomalies. Additionally, it applies five stylized facts of financial returns—volatility clustering, non-Gaussian return distributions, autocorrelation, decreasing autocorrelation in squared returns, and the leverage effect—to provide a comprehensive understanding of Bitcoin's market behavior. This research contributes to the ongoing discourse on cryptocurrency market dynamics, offering insights for risk management and investment strategies in highly volatile environments.

2 Dataset and Methodology

2.1 Methodology

This study adopts a quantitative research methodology. The research focuses on analyzing Bitcoin's price behavior through empirical data, using statistical models to uncover patterns and relationships within the dataset. The goal is to objectively identify trends, outliers, and significant market events.

2.2 Hourly Bitcoin Dataset

This study uses a dataset comprising 44,419 hourly Bitcoin price observations collected between May 15, 2018, and June 9, 2023, sourced from Bitstamp Exchange. Bitstamp is one of the most reputable platforms for Bitcoin trading, ensuring high-quality data for analysis. The dataset includes key variables such as open, high, low, and close prices, in addition to transaction volumes in both Bitcoin and USDT (a stablecoin pegged to the US dollar). These fields enable a rough examination of intraday price trends, volatility patterns, and trading behavior over a five-year period (Table 1).

2.2.1 Statistical Overview and Volatility Characteristics

Bitcoin's price volatility is one of the most distinctive aspects of the dataset. The highest recorded price during the observation period was $69,000 in November 2021, while the lowest was $3,139.76 in December 2018. This substantial price range reflects Bitcoin's sensitivity to market forces, speculative trading, and global economic events.

Table 1 Dataset variables and respective data types

Column name	Data type	Description
Unix timestamp	Bigint	The timestamp in Unix format, which can be converted to local time
Date	Datetime	Timestamp converted to New York Eastern Standard Time (EST)
Symbol	String	The ticker symbol for Bitcoin (BTC)
Open	Float	The opening price for the time period
High	Float	The highest price for the time period
Low	Float	The lowest price for the time period
Close	Float	The closing price for the time period
Volume (Crypto)	Float	The volume transacted in Bitcoin
Volume (Base Ccy)	Float	The volume transacted in USDT (converted base currency)

Table 2 Observed variability of key Bitcoin prices

Column name	Count	Mean	Standard deviation	Minimum	Maximum
Open	44,419	$21,480.56	$16,674.14	$3,139.76	$68,635.05
High	44,419	$21,590.35	$16,768.57	$3,158.34	$69,000.00
Low	44,419	$21,362.77	$16,571.94	$3,122.28	$68,447.00
Close	44,419	$21,481.02	$16,674.19	$3139.76	$68,627.01

- Mean and Median Prices: The average Bitcoin price over the study period was approximately $21,000–$22,000. Price movements were characterized by significant deviations due to speculative market activities and external shocks.
- Price Fluctuations: Especially, price surges occurred during the 2021 bull market, followed by declines in 2022, reflecting Bitcoin's sensitivity to institutional adoption, inflation concerns, and macroeconomic instability.

The dataset's summary statistics, as shown in Table 2, provide key insights into the behavior of Bitcoin's prices across these periods, highlighting the extreme variability in prices.

2.3 *Outlier Detection Using Clustering Analysis*

Outlier detection is crucial in financial data analysis, particularly for highly volatile assets like Bitcoin. Identifying anomalous price movements can reveal insights into market behavior, liquidity, and investor sentiment. In this study, two advanced techniques, t-Distributed Stochastic Neighbor Embedding (t-SNE) and Principal Component Analysis (PCA), were used to detect and analyze outliers in the hourly Bitcoin dataset. Both methods are commonly employed to project high-dimensional data into lower-dimensional spaces, enabling the identification of outliers and providing a better understanding of the underlying price dynamics.

2.3.1 t-Distributed Stochastic Neighbor Embedding (t-SNE)

t-SNE, introduced by van der Maaten and Hinton [16], is a powerful technique for visualizing high-dimensional datasets by mapping them into a low-dimensional space (typically two or three dimensions). This method preserves the local structure of the data, ensuring that data points close in the high-dimensional space remain close in the low-dimensional projection. In this study, t-SNE was applied to the hourly Bitcoin trading data, with each point in the plot representing a specific hourly trading period.

The primary goal of using t-SNE was to cluster similar trading periods together, making it easier to identify potential outliers—trading periods where price behavior deviated significantly from the norm. Outliers identified through t-SNE could correspond to sudden price spikes or crashes that deviate from broader trends. Such outliers might coincide with significant market events, regulatory announcements, or periods of heightened speculative activity, which are common in Bitcoin trading.

However, a limitation of t-SNE is that while it effectively captures local relationships, it may distort global distances between clusters. This limitation was observed in the current study, where some clusters of trading periods appeared close together, despite exhibiting significantly different global behaviors in the high-dimensional space (van der Maaten and Hinton, 2008). To address this limitation and capture the global variance more effectively, Principal Component Analysis (PCA) was employed as a complementary method.

2.3.2 Principal Component Analysis (PCA)

PCA, first introduced by Pearson (1901) and later developed by Hotelling [10], is another widely used dimensionality reduction technique. Unlike t-SNE, PCA focuses on capturing the global structure of the data by identifying the principal components that account for the most variance. PCA transforms the original high-dimensional data into a smaller set of uncorrelated variables (principal components), while preserving the maximum amount of variance in the dataset [12].

In this study, PCA was applied to the hourly Bitcoin dataset to identify the key factors influencing Bitcoin's price movements. The first three principal components (PC1, PC2, and PC3) were plotted in both 2D and 3D visualizations, capturing a significant portion of the variance in the data (Fig. 1). These visualizations helped detect key outliers, such as trading periods with markedly different price, high/low, and volume combinations compared to the rest of the dataset. Notably, the onset of the

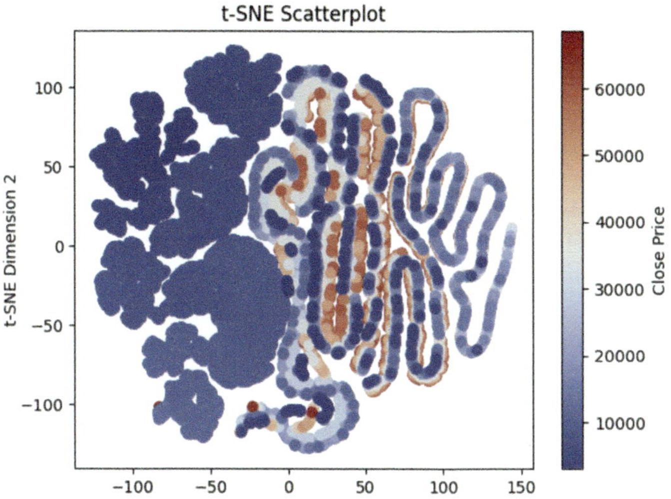

Fig. 1 T-SNE scatterplot for hourly bitcoin price

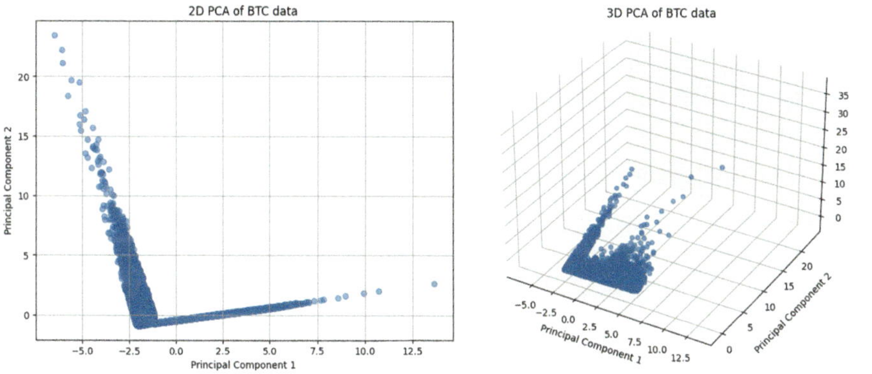

Fig. 2 Two-dimensional PCA of hourly bitcoin data (left) and three-dimensional PCA of hourly bitcoin data (right)

COVID-19 pandemic was identified as a major outlier event, significantly affecting Bitcoin's valuation. PCA's ability to preserve global variance made it well-suited for uncovering such critical market shifts [12] (Fig. 2).

2.3.3 Discussion

The combination of t-SNE and PCA provided a comprehensive approach to analyzing Bitcoin's price dynamics. t-SNE proved highly effective in detecting local relationships between trading periods and identifying anomalous periods in Bitcoin's trading history. Meanwhile, PCA excelled at capturing the broader global structure, making it ideal for understanding significant outliers linked to external market events, such as the COVID-19 pandemic. Together, these methods revealed that Bitcoin's price movements are prone to sudden and extreme fluctuations, often driven by external market shocks, major economic events, or shifts in investor sentiment [5].

By using both t-SNE and PCA, this study demonstrated the importance of applying multiple techniques for outlier detection in financial datasets. The combined approach highlighted Bitcoin's susceptibility to abrupt market changes and offered valuable insights for better understanding and predicting its volatile price behavior.

2.4 Analyzing Hourly Bitcoin Data: Stylized Facts and Visualizations

Under this section, the analysis of the hourly Bitcoin dataset focused on visualizing several key stylized facts, which are recurring empirical patterns in financial data. These facts are essential for interpreting Bitcoin's behavior and making informed predictions about future price movements. The stylized facts examined in this study include non-Gaussian distribution, volatility clustering, autocorrelation, time-varying volatility, and the leverage effect [4, 13]. Recognizing these patterns is vital for understanding Bitcoin's volatility and the associated risks and returns (Fig. 3).

Fact 1: Non-Gaussian distribution of returns

The distribution of Bitcoin returns exhibits fat tails, indicating a higher frequency of extreme events compared to a normal distribution [14]. Analyzing logarithmic returns through histograms and Q-Q plots reveals centrality around zero but heavy-tailed

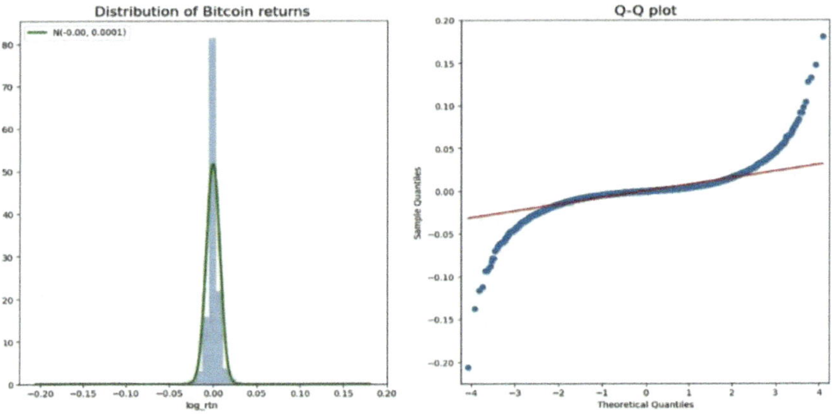

Fig. 3 Presents on the left histogram and right Q-Q plot of Bitcoin returns

behavior, implying increased kurtosis and associated risks. Similar observations in stock markets underscore the likelihood of extreme events, necessitating robust risk management strategies [9]. Additionally, fat tails suggest elevated Bitcoin return volatility, demanding investor readiness for high volatility periods and potential substantial losses [14].

Fact 2: Volatility Clustering

Bitcoin is a volatile asset, with periods of high returns and other periods of negative returns. This volatility clustering is a prevalent feature in financial time series data, and it is important for both risk management and option pricing. Bollerslev [2] first identified volatility clustering, where periods of high or low volatility follow similar periods. This phenomenon is also evident in Bitcoin returns, with high volatility periods often followed by similar phases and vice versa. Other factors, such as investor sentiment, market liquidity, and regulatory changes, can also impact volatility. For example, as depicted on Fig. 4, the average duration of a high volatility period is 3 months, while the average duration of a low volatility period is 6 months. The highest volatility period in the past year was in early 2020, when the price of Bitcoin fell by more than 50% in a matter of weeks. However, it is important to note that volatility clustering does not provide a definitive forecast of future volatility. In addition to volatility clustering, Bitcoin returns also exhibit time-variation, with periods of high and low volatility alternating. This variability is influenced by factors like investor sentiment, market liquidity, and regulatory changes, making return predictions challenging. Nonetheless, understanding these influencing factors can assist investors in managing their risk more effectively. Bitcoin volatility clustering and time-variation are important phenomena that investors should be aware of when making investment decisions. By understanding these factors, investors can better manage their risk and make more informed investment decisions.

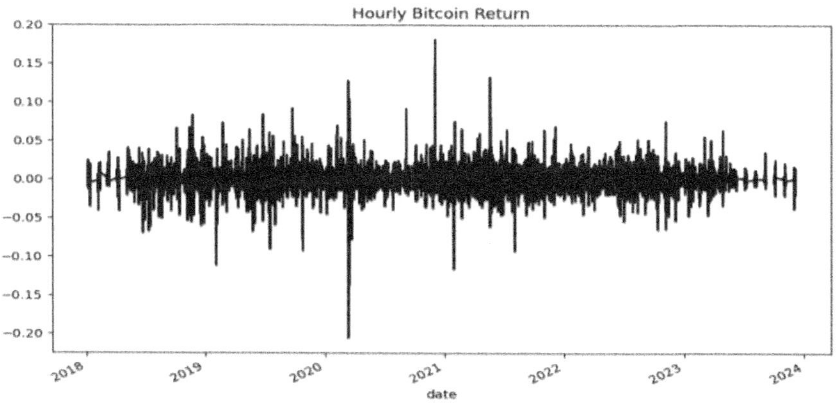

Fig. 4 Illustrates that hourly Bitcoin returns between 2018 and 2023 exhibit volatility clustering upward trend

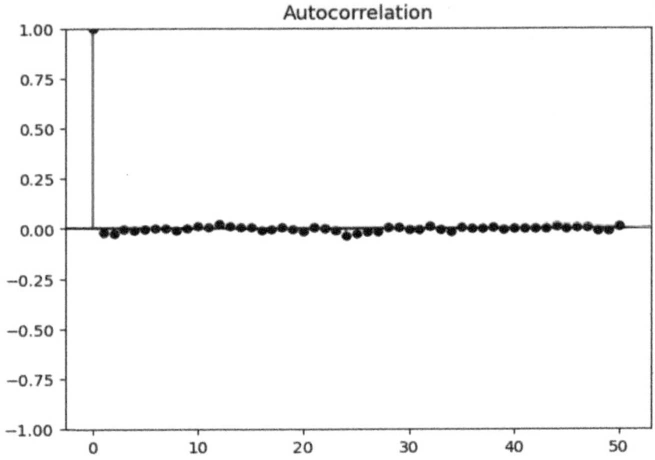

Fig. 5 Shows the autocorrelation function (ACF) plot of Bitcoin returns

Fact 3: Absence of Autocorrelation in Returns

The Autocorrelation Function (ACF) plot, commonly analyzed up to 50 lags, offers insights into data patterns [4]. For Bitcoin, the ACF plot displays negligible autocorrelation beyond lag zero, indicating Bitcoin returns behave akin to a random walk. This aligns with the Efficient Market Hypothesis's weak form, suggesting that all publicly available information is already reflected in prices [9]. Consequently, technical analysis and momentum strategies, reliant on historical data, may have limited effectiveness with Bitcoin [11]. Nevertheless, Bitcoin prices remain influenced by factors beyond past returns, including news events, regulations, and investor sentiment [3, 6, 7]. In conclusion, Bitcoin's ACF analysis (see Fig. 5) indicates minimal impact of past returns on future ones, emphasizing the importance of considering various factors in investment decisions (Fig. 6).

Fact 4: Small and Decreasing Autocorrelation in Squared/Absolute Returns

Autocorrelation plots of Bitcoin's squared and absolute logarithmic returns reveal the presence of volatility clustering, a well-documented characteristic in financial time series data [1, 5, 14]. This phenomenon, where periods of high volatility follow similar periods and vice versa, challenges the suitability of simplistic models like random walks or white noise. Instead, more sophisticated models such as ARCH (Autoregressive Conditional Heteroskedasticity) and GARCH (Generalized Autoregressive Conditional Heteroskedasticity) [2, 8] are better equipped to capture the time-dependent nature of volatility, improving predictive accuracy. These findings underscore the substantial influence of volatility clustering on Bitcoin prices, emphasizing the necessity of advanced models for accurate forecasting.

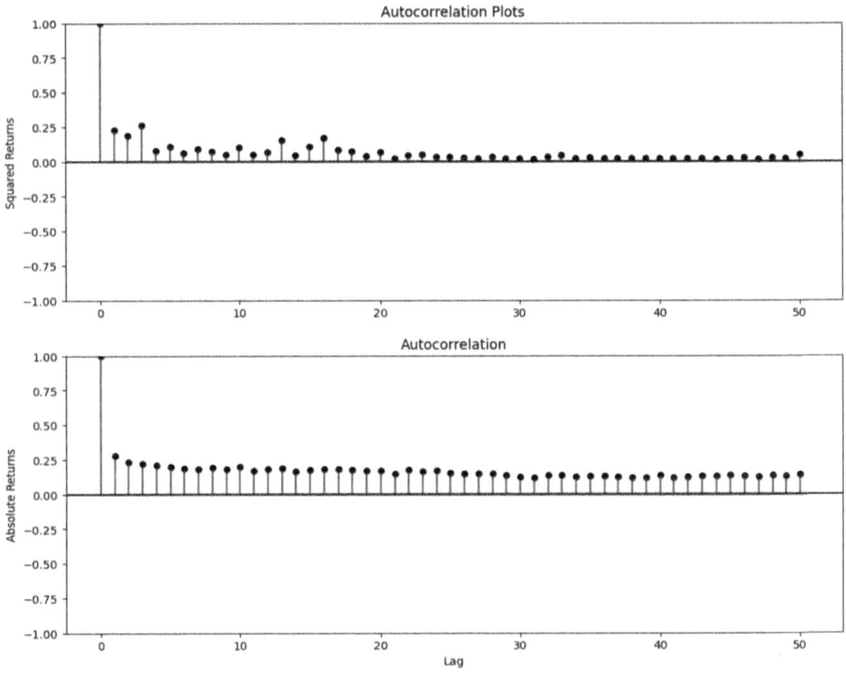

Fig. 6 Illustrates two distinct autocorrelation plots Bitcoin's squared and absolute logarithmic returns respectively

Fact 5: Leverage effect

Bitcoin prices are highly volatile, with significant fluctuations in closing prices, as evidenced on Fig. 7s, both short (21 days) and long (252 days) periods of time. Over a 21-day period, Bitcoin prices experience increases and declines, emphasizing the inherent unpredictability of Bitcoin investment. Logarithmic returns, representing percentage changes in investment, vividly display the substantial variations in Bitcoin returns over a 21-day period. It's noteworthy that periods of substantial positive returns are often followed by equally significant negative returns, emphasizing the uncertain nature of Bitcoin's investment prospects. Over a 252-day period (one year), Bitcoin prices experience significant fluctuations. The 252-day moving volatility, a common metric used to measure the long-term volatility of Bitcoin prices, indicates that increasing values of this indicator indicate heightened price oscillations, magnifying the risks associated with investments.

The chart as explained on Fig. 8 depicts Bitcoin's closing price over time, with moving averages overlaid. This technical indicator helps identify trends and potential reversal points in a market. Since November 2021, Bitcoin's price has been on a downward trend, reaching an all-time high of nearly $69,000. The price has fallen by over 50% until June 2023. The graph shows the highest price, lowest price, and average price, with the black line representing the average price. The yellow line

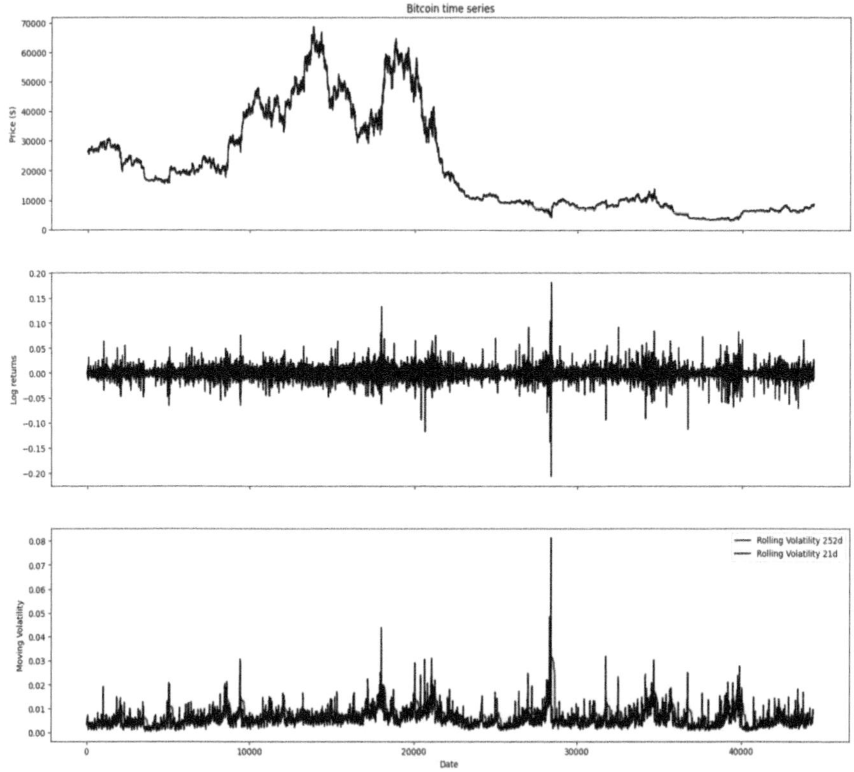

Fig. 7 Presents from top to down: the Bitcoin time series, logarithmic returns, and moving volatility over 21 and 252 days

represents the average price for the past 21 days, the blue line for the past 50 days, and the red line for the past 200 days. The graph also shows an outlier, the candle at the top, representing the all-time high price.

Conclusively, Bitcoin prices are highly volatile, even over both short (21 days) and long (252 days) periods of time. This volatility is evident in the substantial variations in Bitcoin returns and the prevalence of outliers in Bitcoin price data.

2.5 *Discussion*

The analysis of hourly Bitcoin data highlights several critical patterns, or stylized facts, that are fundamental to understanding the cryptocurrency's price behavior. One of the most significant findings is the non-Gaussian distribution of Bitcoin returns, characterized by fat tails. This indicates a higher likelihood of extreme price changes, posing heightened risks for investors. Unlike traditional financial assets that follow

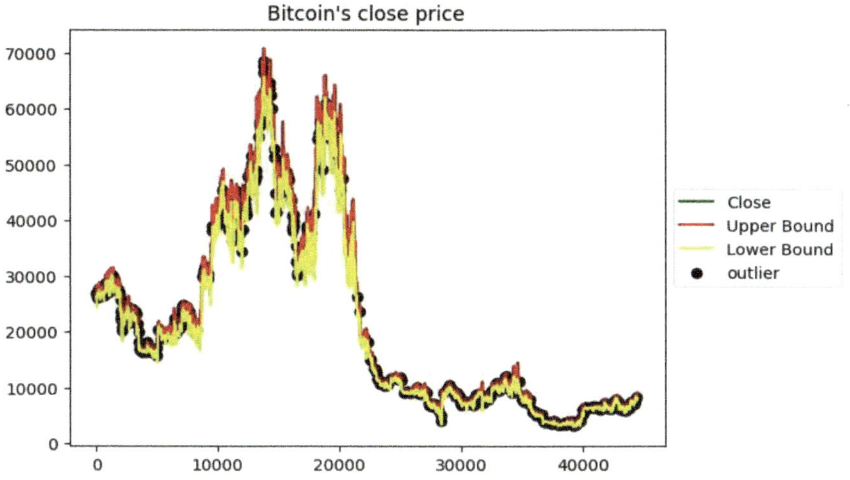

Fig. 8 Demonstrates the graph of outliers' detection using rolling statistics

a normal distribution, Bitcoin's returns are more prone to extreme highs and lows, necessitating robust risk management strategies to protect against substantial losses.

Another important observation is volatility clustering, where periods of high or low volatility tend to follow similar periods. This phenomenon suggests that Bitcoin's market behavior is cyclical, with phases of intense price fluctuations followed by relative calm. While volatility clustering doesn't predict future volatility, it is essential for risk management, especially in a market as unpredictable as Bitcoin. The presence of small but persistent autocorrelation in squared and absolute returns further underscores the need for advanced models like ARCH and GARCH to effectively capture the time-varying nature of Bitcoin's volatility.

Lastly, the leverage effect in Bitcoin prices shows that negative shocks tend to be followed by increased volatility, adding to the asset's inherent unpredictability. The substantial variations in Bitcoin's short- and long-term prices, combined with the prevalence of outliers, emphasize the importance of cautious investment strategies. Overall, Bitcoin's volatility, fat tails, and clustering behavior make it a high-risk asset, and investors must account for these factors when making decisions in this volatile market.

3 Result

3.1 Non-Gaussian Returns and Extreme Events

Tail Risks: The fat tails in Bitcoin's return distribution indicate a higher probability of experiencing extreme price movements, both positive and negative. This implies that large, unexpected gains or losses are more likely to occur compared to traditional assets like stocks or bonds.

Impact on Risk Management: Traditional risk measures like Value at Risk (VaR), which assume a normal distribution, may underestimate the likelihood of extreme events in the Bitcoin market. Investors and risk managers need to employ more robust methods, such as Extreme Value Theory (EVT), to accurately assess and manage tail risks.

3.2 Volatility Clustering and Forecasting Challenges

Persistence of Volatility: The clustering of volatility in Bitcoin's price suggests that periods of high price fluctuations tend to persist. This makes it difficult to predict future price movements, as past volatility can provide clues about potential future volatility.

Implications for Trading and Hedging: Volatility clustering can impact trading strategies, as it may be more profitable to trade during periods of high volatility. Additionally, hedging strategies need to be adjusted to account for the persistence of volatility.

3.3 Absence of Autocorrelation in Returns

Random Walk Behaviour: The lack of autocorrelation in Bitcoin's returns suggests that past price movements have limited predictive power for future returns. This implies that technical analysis methods, which rely on historical price patterns, may be less effective in forecasting Bitcoin's price.

Market Efficiency: The absence of autocorrelation could be interpreted as evidence of weak-form market efficiency, where all publicly available information is already reflected in the price. However, it's important to note that market efficiency is a complex concept, and there may be other factors influencing Bitcoin's price.

3.4 *Small and Decreasing Autocorrelation in Squared/Absolute Returns*

Time-Varying Volatility: The autocorrelation in squared/absolute returns indicates that Bitcoin's volatility is not constant but changes over time. This is consistent with the volatility clustering observed in the data.

ARCH/GARCH Models: Models like Autoregressive Conditional Heteroscedasticity (ARCH) and Generalized Autoregressive Conditional Heteroscedasticity (GARCH) can be used to capture the time-varying nature of Bitcoin's volatility. These models can provide valuable insights into the risk profile of Bitcoin and aid in risk management.

3.5 *Leverage Effect*

Asymmetric Impact of Price Shocks: The leverage effect suggests that negative price shocks have a greater impact on volatility than positive shocks. This asymmetry can increase the risk profile of Bitcoin, particularly during periods of market stress.

Dynamic Risk Management: To effectively manage the risk associated with Bitcoin's leverage effect, investors may need to employ dynamic risk management strategies that can adapt to changing market conditions. This could involve using options or other derivatives to hedge against potential downside risks.

3.6 *Outlier Detection Using Clustering Analysis*

Identifying Significant Events: Outlier detection techniques can help identify significant market events that have a substantial impact on Bitcoin's price. By understanding these events, investors can better assess the factors driving Bitcoin's price and anticipate potential risks.

Risk Assessment and Mitigation: Identifying outliers can also help investors develop strategies to mitigate the risks associated with these events. For example, if an outlier is related to a regulatory announcement, investors may be able to adjust their positions to account for the potential impact.

4 Conclusion and Future Research

This comprehensive analysis of Bitcoin's hourly price data from May 2018 to June 2023 has unveiled several key characteristics of its market dynamics. Bitcoin's returns exhibit a non-normal distribution, characterized by "fat tails" and a peaked centre, indicating a higher likelihood of extreme price movements. Its volatility is clustered, making it challenging to predict future price movements. Past price movements have limited predictive power for future returns, suggesting that technical analysis may be less effective. The volatility of Bitcoin's returns is not constant and shows a pattern of clustering, suggesting the use of models like ARCH and GARCH. Bitcoin exhibits a leverage effect, where negative price shocks are often followed by increased volatility. Outlier analysis revealed significant anomalies in Bitcoin's price behaviour, often corresponding to major market events.

To conclude, these findings highlight the volatile and unpredictable nature of Bitcoin, emphasizing the need for robust risk management strategies. Investors and traders operating in the Bitcoin market should be aware of the potential for extreme price movements, the clustering of volatility, and the limited effectiveness of traditional technical analysis.

Henceforth, future research could expand upon these findings by comparing Bitcoin's market dynamics to other cryptocurrencies, employing more sophisticated machine learning techniques, investigating Bitcoin's price behaviour across different time frames, investigating the impact of macroeconomic factors, and examining the influence of regulatory changes and technological advancements. By pursuing these research avenues, we can gain a more comprehensive understanding of Bitcoin's role in the financial landscape and develop more effective strategies for investing and risk management in the cryptocurrency market.

References

1. Black, F. "Studies of Stock Market Volatility Changes." *Proceedings of the American Statistical Association*. (1976)
2. Bollerslev, T. "Generalized Autoregressive Conditional Heteroskedasticity." *Journal of Econometrics*. (1986)
3. Bouri, E., Gupta, R., Lau, C. K. M., Roubaud, D., Wang, S. "Bitcoin and global financial stress: A copula-based approach to dependence and causality in the quantiles." *The Quarterly Review of Economics and Finance*, 69, 297-307 (2017)
4. Box, G. E. P., Jenkins, G. M., Reinsel, G. C.*Time Series Analysis: Forecasting and Control* (4th ed.). Wiley. ISBN: 978-1-118-61919-3 (2013)
5. Cont, R. "Empirical Properties of Asset Returns: Stylized Facts and Statistical Issues." *Quantitative Finance*. (2001)
6. Corbet, S., Lucey, B. M., & Yarovaya, L. "Cryptocurrencies as a financial asset: A systematic analysis." *International Review of Financial Analysis*, 62, 182-1992. (2018)
7. Dyhrberg, A. H. "Hedging capabilities of Bitcoin: Is it the virtual gold?" *Finance Research Letters*, 16, 139-1441. (2016)
8. Engle, R. F. (1982). "Autoregressive Conditional Heteroscedasticity with Estimates of the Variance of United Kingdom Inflation." *Econometrica*, 50(4), 987-1007

9. Fama, E. F. "Efficient Capital Markets: A Review of Theory and Empirical Work." The Journal of Finance. (1970)
10. Hotelling, H. "Analysis of a complex of statistical variables into principal components." *Journal of Educational Psychology*, 24(6), 417–441. (1933)
11. Jegadeesh, N., Titman, S. "Returns to Buying Winners and Selling Losers: Implications for Stock Market Efficiency." The Journal of Finance. (1993)
12. Jolliffe, I. T., Cadima, J.. "Principal Component Analysis: A Review and Recent Developments." Philosophical Transactions of the Royal Society A. (2016)
13. Lewinson, R. "Bitcoin's Extreme Price Movements: Evidence from Fat-Tailed Return Distributions." *Financial Review*. (2022)
14. Mandelbrot, B. "The Variation of Certain Speculative Prices." *The Journal of Business*. (1963)
15. Pearson, K. "On Lines and Planes of Closest Fit to Systems of Points in Space." *Philosophical Magazine*, Series 6, 2(11), 559–572 . (1901)
16. van der Maaten, L., Hinton, G. (2008). "Visualizing Data Using t-SNE." Journal of Machine Learning Research

Unlocking the Potential of Intelligent Blockchain: A Quantitative Analysis

Vincenza Carchiolo and Michele Malgeri

Abstract This paper presents a quantitative analysis of research papers related to Intelligent Blockchain. Intelligent blockchain is an emerging topic that aims to study the impact of artificial intelligence into blockchain technology and its application. The goal of this paper is twofold: to provide a quantitative analysis and to experiment with new methodologies for studying large, significant topics. The first objective is to provide a valuable analysis of the corpus of publications using various metrics, such as bibliometric indices integrated with techniques from complex network analysis, to evaluate how these publications have been received by the scientific community. The second objective is to validate the effectiveness of network analysis in discovering hidden relationships between researchers and countries, thereby enhancing our understanding of certain dynamics within the scientific community.

Keywords Intelligent blockchain · Complex networks · Quantitative analysis

1 Introduction

In the dynamic realm of technology, the fusion of innovative concepts often sparks transformative advancements. The convergence of artificial intelligence (AI) and blockchain technology stands as a prime example of this synergistic potential. Intelligent Blockchain (IB), a burgeoning field at the intersection of these two powerful domains, holds immense promise for revolutionizing various aspects of our digital landscape. Blockchain, as widely known, is a decentralized and distributed technology that allows transactions to be recorded in a secure, transparent, and immutable manner. Its applications are numerous. This technology is based on a graph model where each node represents a computer that participates in the validation and recording of transactions. Transactions are grouped into blocks, each containing a set of

V. Carchiolo (✉) · M. Malgeri
Dipartimento di Ingegneria Elettrica, Elettronica e Informatica, Università degli Studi di Catania, Catania, Italy
e-mail: vincenza.carchiolo@unict.it

M. Malgeri
e-mail: michele.malgeri@unict.it

N. Polatidis et al. (eds.), *Intelligent Distributed Computing XVII*, Studies in Computational Intelligence 1203, https://doi.org/10.1007/978-3-031-87639-4_13

transactions and a unique hash that links it to the previous block, thus forming a continuous and secure chain. Blockchain uses cryptographic algorithms to ensure that data cannot be altered once recorded, making it immutable. The decentralization of the network implies that no single entity has complete control, enhancing the security and transparency of operations. Blockchain originated as the foundation for Bitcoin, but its applications have extended to many other sectors, such as finance, supply chain management, smart contracts, and healthcare.The security of blockchain is further reinforced by consensus mechanisms like Proof of Work and Proof of Stake, which require the collaboration of multiple nodes to validate transactions.

AI, with its ability to process vast amounts of data, learn from patterns, and make intelligent decisions, offers a compelling solution to some of the limitations of blockchain like scalability, adaptability and autonomy. The integration of AI into blockchain, often referred as Intelligent Blockchain, has the potential to revolutionize various industries and domains, including Financial Services [17], Supply Chain Management [9], Healthcare [2] and Governance [31].

This work aims at analyzing the trend of publications dealing with Intelligent Blockchain using a quantitative approach in order to obtain a general overview of the scientific output. A similar approach to analyze publication data is used in [5]. Therefore, the objective is twofold: to highlight the most important challenges in the field, providing a guide to researchers, and to experiment with the technique used to analyze data in the field of evaluating the quality of scientific outcomes. To perform a significant analysis, we need a large quantity of bibliographic data. For this reason, we used Elsevier's Scopus database, one of the largest databases of scientific publications, executing several queries tailored to extract data related to Intelligent Blockchain. The contribution of this paper lies in both method and results themselves. The method, a systematic literature review, and the way data has been explored could be easily extended to several other fields. The results offer different views, such as the level of engagement and collaboration among researchers in the field, identifying leading researchers and institutions, the volume of research output, and the nature of collaborative efforts across various disciplines and geographical areas. Finally, we provide a report on the main themes and topics addressed in recent research papers.

Section 2 introduces the method used to conduct our investigation. Section 3 introduces the dataset creation and its bibliometric analysis, in Sect. 4 the result of network based analisys are presented and in Sect. 5 the thematic analysis. Finally, in Sect. 6 some conclusive remarks are discussed.

2 Review Methods

In this paper, we conducted a comprehensive quantitative analysis of the topic. Quantitative analysis encompasses a range of methods that use numerical data to assess and understand phenomena. These methods can be broadly categorized into three main types: bibliometric analysis, network analysis, and thematic analysis. There are numerous systematic reviews in the literature on the topic [28], but they are often limited to bibliometric aspects or specific applications of blockchain in certain sectors [1, 8, 25].

2.1 Bibliometric Analysis

The first, and perhaps the most classic, type of quantitative analysis is bibliometric analysis. Bibliometric analysis focuses on the quantitative examination of published works, such as scientific articles, books, and patents. It provides insights into the growth, authorship patterns, and citation impact of a particular field of research. By analyzing bibliometric data, researchers can identify trends in research topics, collaborations, and institutional contributions. Bibliometric analysis provides valuable insights into the growth, structure, and impact of research in Intelligent Blockchain. By analyzing publication volume, geographical distribution, citation patterns, and publication venues, we can gain a comprehensive understanding of the field's evolution, key contributors, and emerging trends. This information can inform future research directions, identify areas for collaboration, and contribute to the broader understanding of Intelligent Blockchain.

In this section, we employ bibliometric analysis to investigate the research landscape of Intelligent Blockchain. To assess the evolution of research interest in Intelligent Blockchain, we analyze the publication volume over time (*Publication Volume Analysis*). This involves tracking the number of publications related to Intelligent Blockchain published each year. An increasing publication volume indicates growing interest and activity in the field. To understand the geographical distribution of research in Intelligent Blockchain, we analyze the affiliations of the authors of the publications (*Geographical Distribution Analysis*). This involves identifying the countries of affiliation for all authors and counting the number of publications associated with each country. This analysis reveals the geographical distribution of research activity and can identify centers of excellence in the field. *Citation analysis* examines the frequency with which publications are cited by other researchers. A high citation count indicates that a publication is influential and has made a significant contribution to the field. To account for the time since publication, we consider the citation rate, which is the average number of citations per year received by a publication. Finally, *Publication Venue Analysis* examines the types of venues where Intelligent Blockchain research is published. This involves identifying the type of publication (e.g., journal article, conference paper) for each publication. This analysis can reveal the preferred venues for disseminating research in the field and provide insights into the types of contributions that are considered most valuable.

The selection of the set of papers to be analyzed is crucial for the validity and reliability of bibliometric analysis. In this study, we carefully crafted a query on the Scopus database [3, 11, 12] to capture relevant publications related to Intelligent Blockchain. The query was designed to be comprehensive and accurate, ensuring that the selected papers accurately represent the research landscape in the field.

2.2 Network Analysis

The second approach, network analysis, investigates the relationships and interactions among entities within a network, such as individuals, institutions, publications, and concepts. By examining the network structure, researchers can identify influential actors, collaborative patterns, and knowledge diffusion pathways. The proposed method employs complex networks to represent the relationships between publications, authors, and other entities, facilitating the identification of research clusters, key authors, and knowledge flows. There are a lot of example in literature delving with of co-authors networks [4, 22]. In this paper, we will perform an analysis on the interconnection among affiliations (i.e. name of the research institution, company, or other organization where the authors conduct their research or work) where the authors work to discover how collaborations between different entities are organized. We refer to this network as a **Co-Affiliation Network** (CANet), where nodes represent the names of affiliations A_i, and an edge $E(A_1, A\_2)$ exists when there is at least one paper whose authors belong to both A_1 and A_2. The analysis conducted on this network aims to demonstrate the degree of cooperation among various research entities regarding the topic.

To analyze the degree of collaboration, we evaluate the degree of each node of CANet. In graph theory, the degree of a node is the number of edges incident to that node that represents the number of connections the node has with other nodes in the graph. Essentially, it indicates the level of collaboration of each node A_i with others A_j Nodes with a high degree may represent influential affiliations, attracting researchers from other affiliations.

To gain a deeper understanding of collaboration patterns in Intelligent Block-chain research, we examine the node degree also known as structural degree. By examining the evolution of node degree over time and across different nations, we can identify patterns of collaboration and potential disparities in research practices. This analysis involves comparing the average node degree, the distribution of node degrees, and the evolution of node degree over time for each nation. By identifying similarities and differences in collaboration patterns, we can gain insights into the factors that influence research practices and collaboration dynamics. In fact, the analysis of node degree can reveal hidden collaboration patterns that may be influenced by factors such as geographical proximity, cultural norms, or institutional affiliations. By identifying nodes with high degree and examining their connections, we can uncover dense clusters of collaboration that may be indicative of shared research interests, common funding sources, or established research partnerships. By focusing on counties that have made significant contributions to Intelligent Blockchain research, we can further investigate state-specific collaboration trends. This involves analyzing the node degree of research institutions within each nation and identifying patterns that may be influenced by national policies, funding mechanisms, or research priorities.

2.3 Thematic Analysis

Thematic analysis is a research method aimed at identifying and interpreting underlying themes and patterns within textual data. It involves a systematic process of coding, categorizing, and interpreting text segments to uncover recurring ideas, concepts, and perspectives. In this study, we use two complementary approaches to thematic analysis: Subject Area Analysis and author keywords based analysis. The subject are extracted from Scopus, the bibliographic database, that organizes publications by associating each paper with a *Subject Area* (SA) derived from the journal or conference where it was published.

By analyzing the distribution of publications across different *Subject Area*, we can identify the diverse applications of Intelligent Blockchain technology across various fields. This analysis provides insights into the breadth and depth of research in this area. In addition to the subject area classification provided by Scopus, we also analyze the author keywords of the most cited papers to highlight the main areas of interest among researchers.

3 Dataset Description and Bibliometric Analysis

The aforementioned dataset was constructed with the aim of encompassing all literature related to Intelligent Blockchain technology.

To achieve this, a query was executed on Scopus, in May 2024 retrieving 11 524 papers, which contains approximately 82 million publications, to select all articles using keywords, titles, and abstracts containing the both the term "Blockchain" and at least one of the following: "Artificial Intelligence", "Machine Learning", "Deep Learning", "Artificial Neural Networks", "Machine Learning Algorithms", "Cognitive Robotics", "General Artificial Intelligence", "Applied Artificial Intelligence", "Computer Vision", "Natural Language Processing", "Expert Systems", "Pattern Recognition", "Supervised Learning", "Unsupervised Learning", "Reinforcement Learning", "Convolutional Neural Networks", "Recurrent Neural Networks", "Optimization Algorithms", "Data Analysis", "Artificial Intelligence Applications".

The query was executed. Figure 1 shows the trend over the years from 2015 to 2024 of number of publications, where *Publication Volume Analysis* that shows a steady increase, indicating a growing interest in the combined topics. The observed decrease in 2024 can be attributed to the dataset being limited to the first five months of the year.

The *Geographical Analysis* reveals that the top 10 countries, in terms of the number of publications, contribute to about two-thirds of the total number of publications. Figure 2 illustrates the geographical distribution over the entire period (2015-2024), where the three countries with the highest number of publications are China, India, and the USA, in that order, with a notable predominance of China and India.

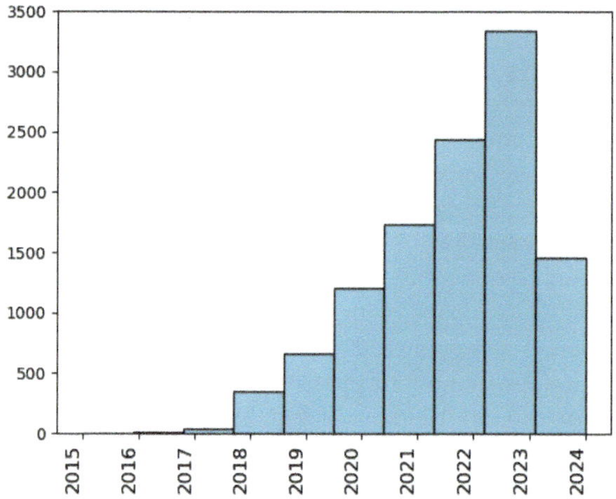

Fig. 1 Trend of publication from 2015 to 2024

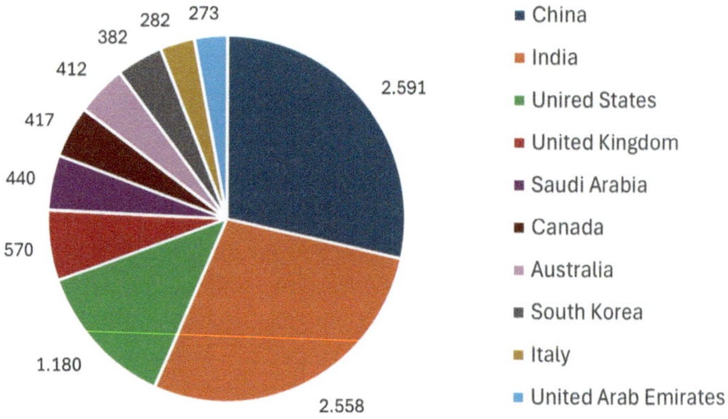

Fig. 2 Distribution of publications over main countries

Figure 3 shows the distribution of publications over the years for each country and indicates that in the peak years 2022 and 2023, India overtakes also China.

Figures 4 and 5 focus on *Citation Analysis*. Figure 4a shows the distribution of citations across all publications on Intelligent Blockchain in the dataset. The x-axis represents the number of citations received by each publication, while the y-axis indicates the number of publications falling within each citation range. The use of a logarithmic scale on the y-axis effectively conveys the distribution of citations, highlighting the prevalence of publications with lower citation counts and the presence of a few highly cited publications. Figure 4b complements the previous analysis by examining the citation rate distribution. In this case, the x-axis represents the citation rate (citations per year) of each publication. Again, the figure highlights that most publications have a low citation rate, while only a few publications exhibit a high

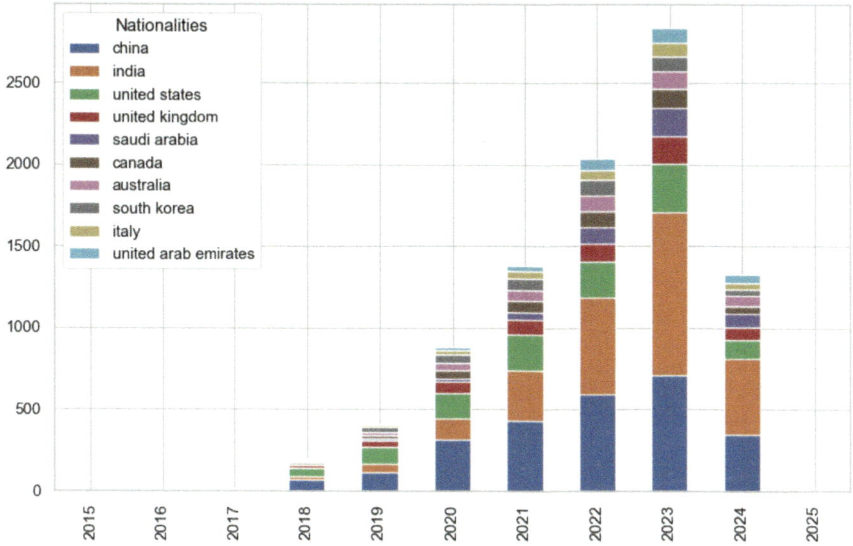

Fig. 3 Trend of publications over main countries and years

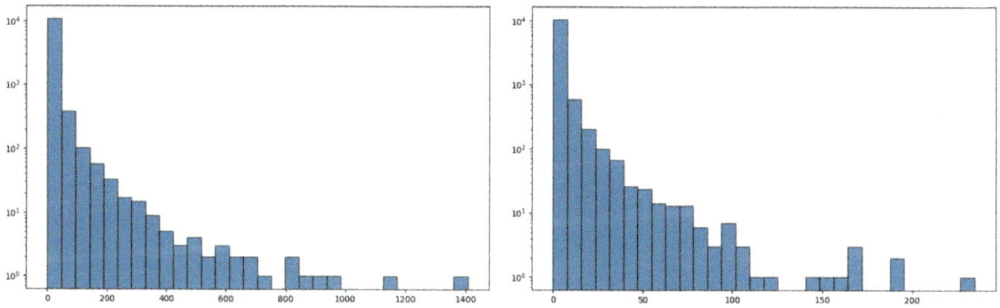

(a) Citation distribution

(b) Rate Citation distribution

Fig. 4 Distribution of citation in absolute and percentage

citation rate. Finally, Fig. 5 offers a detailed overview of the temporal distribution of citations, depicting the spread of article citations across various years. These figures consistently demonstrate an increasing trend in citations for publications within the Intelligent Blockchain domain. This upward trend is likely attributed to the growing interest in this field and the proliferation of research publications. The observed distribution of citations suggests that, while a majority of publications receive a relatively low number of citations, a small subset of publications has gained significant traction within the research community, as evidenced by their high citation counts and citation rates. The citation analysis provides valuable insights into the impact and dissemination of research within the Intelligent Blockchain domain. The observed trends highlight the growing significance of this field and the emergence of highly influential publications that are shaping the trajectory of research in this area.

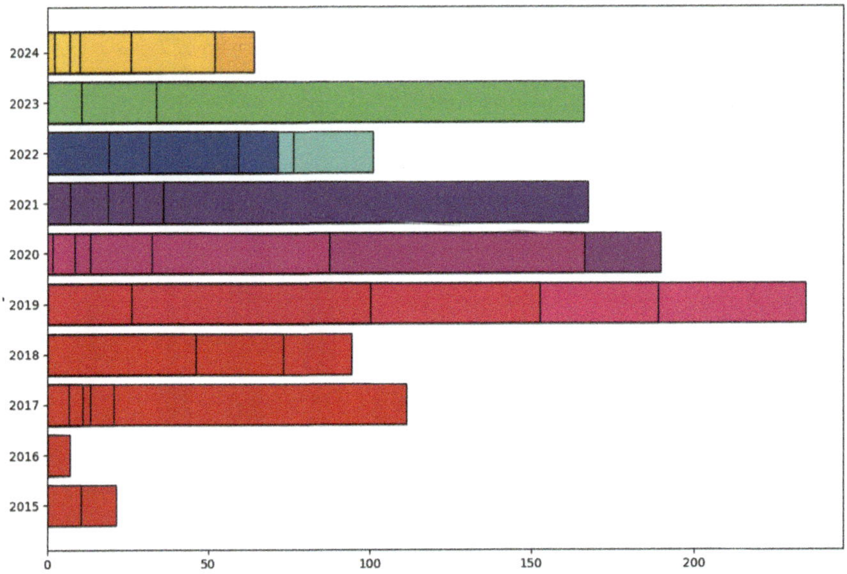

Fig. 5 Percentage of Citation evolution (x-axis). Each year (y-axis), is depicted by a horizontal bar, with the length of the bar representing the percentage of citations received by publications in that year

Publication Venue Analysis uses the Scopus information on the publication type of articles. The two primary publication types are "Article" and "Conference Paper," each accounting for over 30% of the total publications. The distribution of articles Fig. 6 reveals a broad spread across a vast number of journals. The extensive diversity of publication venues reflects the broad interest in Intelligent Blockchain research. While this trend indicates a general openness to publishing in various journals, it prompts an investigating into whether specific journals are preferred by researchers. This preference could be attributed to factors such as faster review processes or higher journal impact factors. Similar to the distribution of articles, conference papers Fig. 6 are also dispersed across a wide range of conferences. This observation highlights the significant interest in presenting research findings at conferences within the Intelligent Blockchain domain. The diversity of conference venues suggests that researchers are actively engaging in disseminating their work across various specialized conferences, potentially targeting specific research communities or subfields within the broader Intelligent Blockchain landscape.

The observed distribution of publication types and the diversity of publication venues underscore the importance of carefully selecting the dataset for analysis. A comprehensive dataset should encompass both articles and conference papers, ensuring that the analysis reflects the breadth of research contributions within the Intelligent Blockchain domain. Additionally, consideration should be given to the

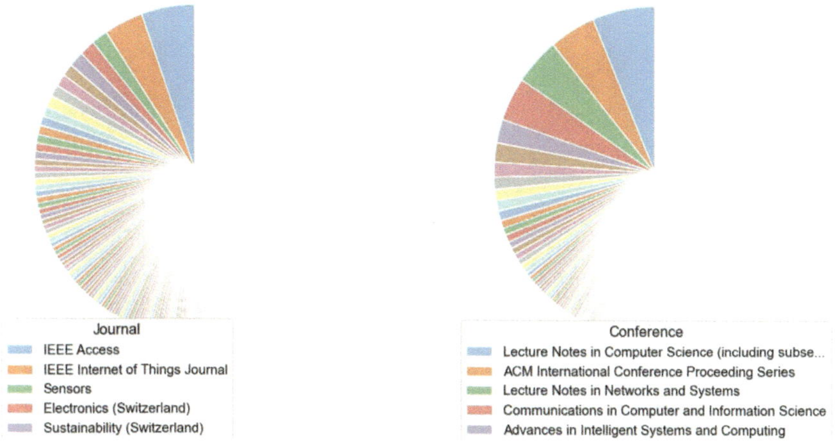

Fig. 6 Distribution of publication over type of publications (journal or conference)

Table 1 Network topology figures

Notation	Description	Value
$\|V\|$	Number of vertices	19,301
$\|E\|$	Number of edges	40,420
σ	Density	0.5
k	Average degree	4.19
N_c	Number of connected components	4,828
$\|lcc\|$	Number of nodes in the largest connected component	1,301

wide range of journals and conferences involved in publishing Intelligent Blockchain research.

4 Co-Affiliation Network

This section details the analysis of CANet. We extracted the authors' affiliations from the dataset, which comprises over 11,000 papers, and constructed the corresponding co-affiliation network. The key characteristics of this network are summarized in Table 1. As illustrated in Fig. 7, the network consists of numerous small connected components, also highlighted by low density (σ) and high fragmentation.

Based on the degree distribution (refer to Fig. 8), a power-law distribution is evident, with some deviations. Notably, 125 out of 19 301 nodes exhibit degree

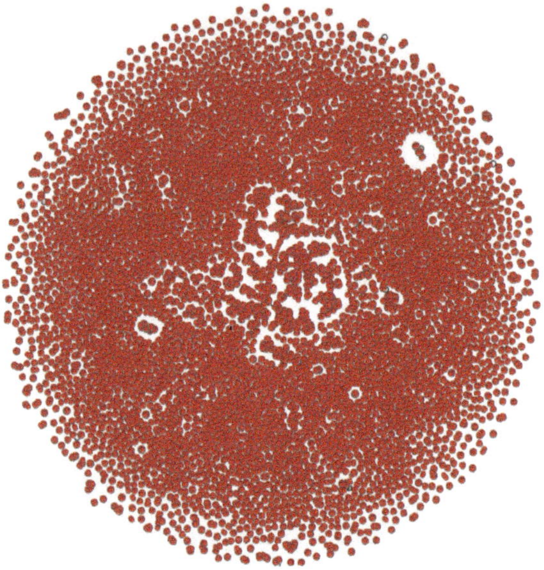

Fig. 7 Co-affiliation network

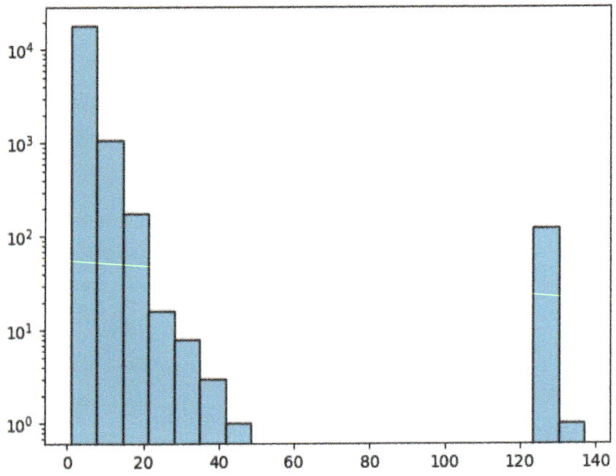

Fig. 8 Degree distribution, (y-axis is in logarithmic scale)

values surpassing 120. These research centers are situated in five countries: Germany, Greece, India, the Netherlands, and the United States.

Next, we discuss the average degree distribution, calculated grouping the CANet nodes by country under various conditions: Figure 9 illustrates the degree distribution across all countries, while Fig. 10 filters out countries with degrees less than 5. In Fig. 11, we consider only those countries where the affiliation degree exceeds 20. Figure 9 reveals a skewed distribution, with only a few countries exhibiting an average node degree exceeding 5, while Fig. 10 provides a closer examination of the countries with an average node degree above 5.

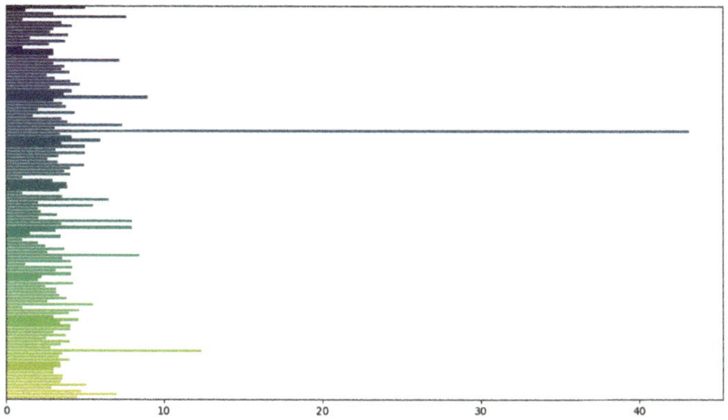

Fig. 9 Distribution of average degree over countries (y-axis)

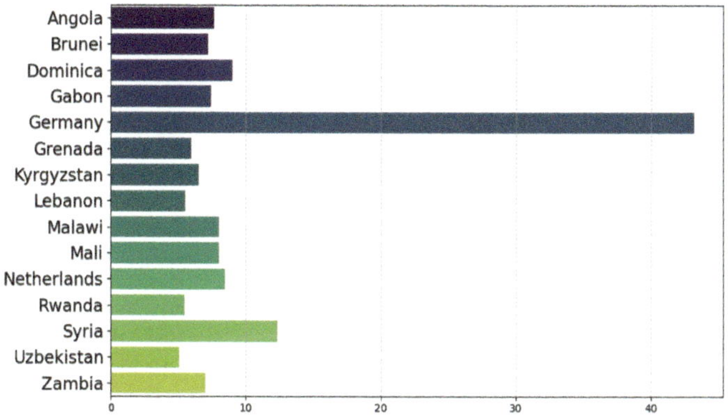

Fig. 10 Average degree distribution of the countries where value is greater than 5

This observation can be attributed to several factors. One possibility is that these countries have a limited number of research institutions that collaborate extensively with each other, leading to a higher average node degree. Alternatively, these countries may have many research institutions, but most of them engage in limited collaboration. However, further investigation is necessary to fully understand the underlying reasons for this distribution. It is noteworthy that, except for Germany, all countries with an average node degree above 5 are relatively small. This suggests that these countries may have a higher propensity to collaborate with institutions from other countries, potentially due to a smaller domestic research landscape. Additionally, these countries tend to have a lower average number of publications within the dataset, which may further contribute to their higher average node degree.

To investigate research structures characterized by extensive collaboration, we focused on CANet nodes with a degree of 20 or higher.

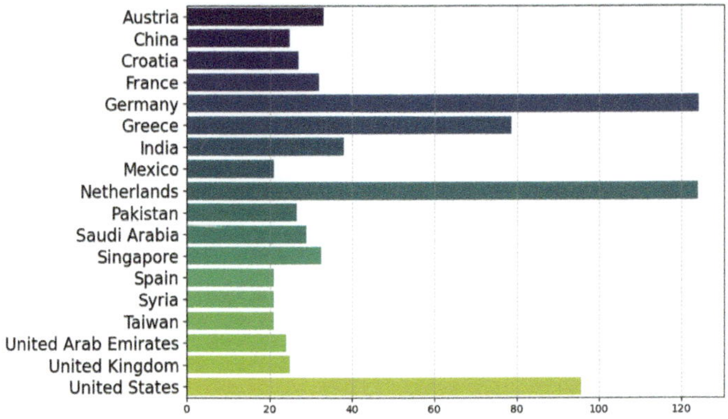

Fig. 11 Average degree distribution for affiliations with degree greater than 20

This approach aims to identify highly connected nodes that serve as significant collaboration hubs within the network. We examined the contribution of these structures to the average node degree in their respective countries. Figure 11 presents the distribution of average node degree across countries, focusing primarily on those with an average node degree above 5 for clarity in the graphical representation.

The analysis reveals that, except for Germany, the countries with high-degree research structures differ from those identified in the previous analysis based on average node degree. However, these countries share the characteristic of having a relatively large average number of publications in the dataset. This observation suggests that in Germany, research structures tend to exhibit higher degrees of collaboration on average. In contrast, the other countries with high-degree structures appear to have a mix of highly collaborative hubs and smaller structures with fewer collaborations. This diversity in collaboration patterns may contribute to the variation in average node degree across these countries. The analysis of high-degree research structures offers deeper insights into the collaborative dynamics within Intelligent Blockchain research. The observed variations in collaboration patterns among countries underscore the importance of examining both national-level and structural-level factors in analyzing collaboration networks.

5 Subject and Keywords Analysis

5.1 Subject Area Analysis

This analysis represents a preliminary attempt to study the subject, focusing on a cumulative analysis of publications across the ten most relevant subject areas, revealing a consistent growth in scientific article numbers.

According to the subject area classification by Scopus, Table 2 contains the num-

Table 2 Number of documents for each Subject area (ref. column is the official acronym in Scopus)

Ref.	Subject area name	Documents
CS	Computer science	8756
ENG	Engineering	4956
MAT	Mathematics	3205
DS	Decision sciences	1960
BMA	Business management and accounting	1246
SS	Social sciences	1069
EN	Energy	707
PA	Physics and astronomy	699
MED	Medicine	694
EEF	Economics econometrics and finance	603
MS	Materials science	521
ES	Environmental science	403
BGMB	Biochemistry genetics and molecular biology	262
ABS	Agricultural and biological sciences	161
CHE	Chemical engineering	149
CHM	Chemistry	145
AH	Arts and humanities	95
EPS	Earth and planetary sciences	89
HP	Health professions	87
MUL	Multidisciplinary	73
PSY	Psychology	59
NS	Neuroscience	48
PTP	Pharmacology toxicology and pharmaceutics	29
IM	Immunology and microbiology	28
NUR	Nursing	22
DNT	Dentistry	5
VET	Veterinary	0

ber of paper in each subject area. It shows that the majority of publications fall within the fields of Computer Science or Engineering, with Computer Science being the predominant subject area. It also highlights the extensive multidisciplinary applications of blockchain technology (note that only the 'Veterinary' subject area has no occurrences), while the Fig. 12 shows that 10 subject areas account for more than 90% of the publications and Computer Science and Engineering accounts 55%.

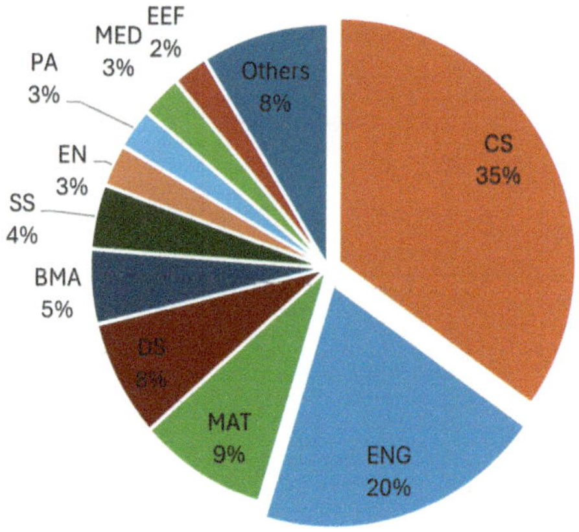

Fig. 12 Distribution of papers over subject areas

Figures 13, 14, and 15 illustrate the increase in article numbers across these 10 subject areas. Given the substantial difference in article counts, we used different figures to arrange the scale. In any case, examining the temporal growth of publications, a clear positive trend emerges over the years, indicative of increasingly intense and continuous research activity.

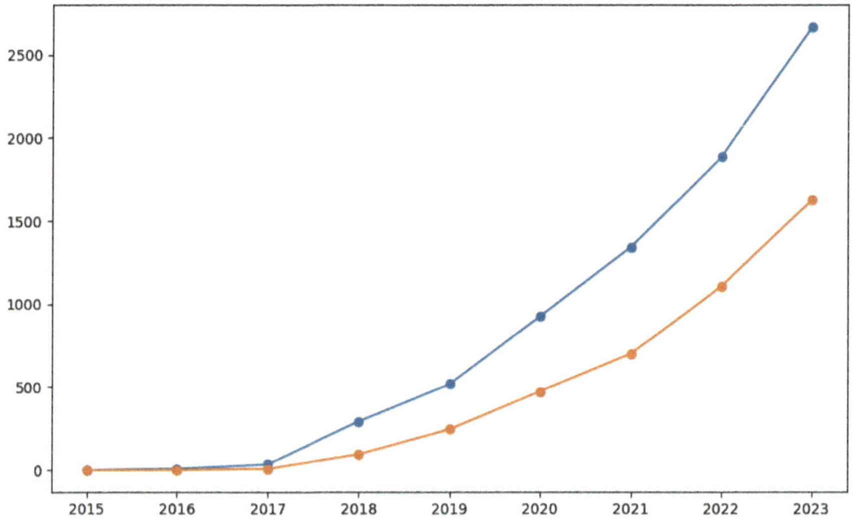

Fig. 13 Subject areas trend. Blu = CS, Orange = ENG

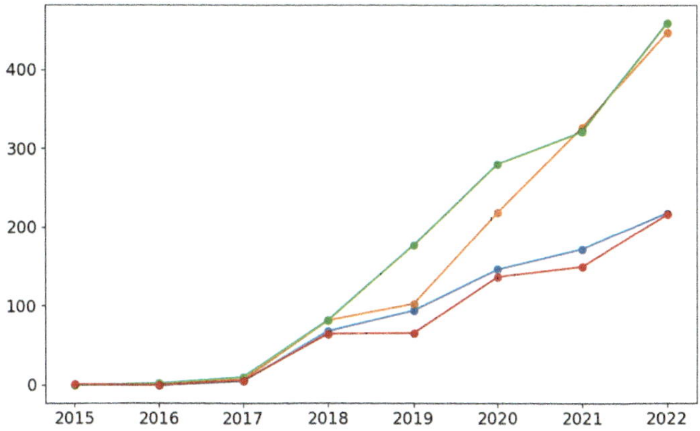

Fig. 14 Subject areas trend. Blu = BMA, Green = MAT, Orange = DS, Red = SS

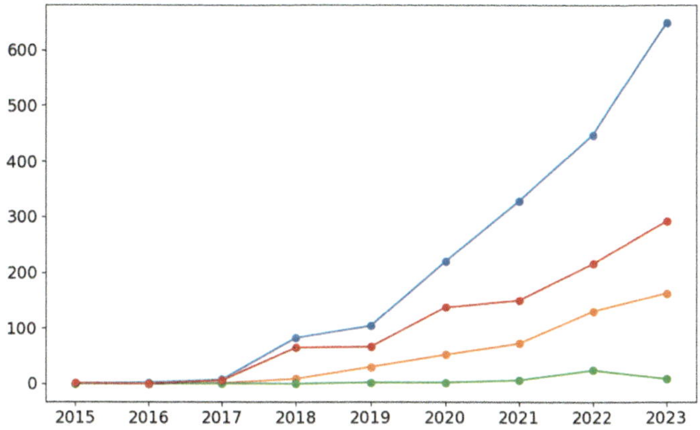

Fig. 15 Subject areas trend. Blu = EEF, Green = MED, Orange = EN, Red = PA

5.2 *Most Cited Paper*

In this subsection, we present a summary of the most cited papers within the dataset, taking into account those with at least 500 citations. High number of citation is one of the metrics that suggests that these papers represent significant contributions to the field of Intelligent Blockchain and provide valuable insights into the most impactful research themes.

Table 3 provides an overview of 16 selected research papers, each exploring the application of blockchain technology in various domains. The table is structured with two columns: the former provides the reference in the bibliography, and the latter the application domain of blockchain technology.

Table 3 Overview of most cited paper related to their application domain

Reference	Research and application domain field
[18]	5G and 6G communication
[10]	Covid-19, Digital transformation
[26]	Covid-19, Digital transformation
[23]	Healthcare, Internet of things
[24]	Internet of things
[7]	5G and 6G communication
[6]	Covid-19, Internet of Things
[19]	Security and privacy
[21]	Digital agriculture
[30]	6G communication
[27]	Digital transformation, Internet of thing
[16]	Internet of thing
[13]	Food supply chain
[15]	Healthcare, Data Sharing
[29]	Food quality;
[14]	Healthcare, Security and privacy
[20]	Secure protocol

It is noteworthy that the adopted classification is more specific than the traditional Subject Areas typically assigned to publications. This granular approach allows for more precise identification of the thematic areas addressed in each paper. Observing the table, we can identify several application domains for blockchain technology, including: Security and Privacy ([19]), 5G and 6G Communications ([7, 18, 30]), Internet of Things (IoT) ([16, 23, 24, 27]), Secure Protocols ([20]). Moreover, there are several application domains including Healthcare ([6, 14, 15, 23]), Digital Transformation ([10, 26, 27]), Digital Agriculture ([21]), Food Supply Chain ([13]) and Food Quality ([29]) It is interesting to note the presence of research exploring the application of blockchain in response to emerging challenges, such as the Covid-19 pandemic ([6, 10, 26]).

6 Conclusions

The results from a quantitative analysis based on bibliometric and geographic analyses and their correlations have provided valuable insights into the evolution and dissemination of research within a specific field. Integrating bibliometric techniques with geographic data enables researchers to comprehensively understand research trends, collaboration patterns, and the global distribution of knowledge production.

Co-affiliation network analysis (CANet) has emerged as a valuable tool for exploring collaboration patterns and knowledge exchange within a specific field. By examining the connections between individuals or organizations based on their shared affiliations, CANet analysis provides insights into the dynamics of collaboration, the diffusion of ideas, and the emergence of research clusters. All topics require further investigation, such as the utilization of a weighted CANet.

The thematic analysis highlights the application fields of Intelligent Blockchain, but future studies aim to investigate the use of Natural Language Processing for a large-scale automated analysis of author keywords.

Acknowledgements Partially supported by the project UDMA

References

1. Agbo, C.C., Mahmoud, Q.H., Eklund, J.M.: Blockchain technology in healthcare: A systematic review. Healthcare **7**(2), 56 (2019). https://doi.org/10.3390/healthcare7020056
2. Attaran, M.: Blockchain technology in healthcare: Challenges and opportunities. International Journal of Healthcare Management **15**(1), 70–83 (2022). https://doi.org/10.1080/20479700.2020.1843887
3. Baas, J., Schotten, M., Plume, A., Côté, G., Karimi, R.: Scopus as a curated, high-quality bibliometric data source for academic research in quantitative science studies. Quantitative Science Studies **1**(1), 377–386 (02 2020). https://doi.org/10.1162/qss_a_00019
4. Carchiolo, V., Grassia, M., Malgeri, M., Mangioni, G.: Co-authorship networks analysis to discover collaboration patterns among italian researchers. Future Internet **14**(6) (2022)https://doi.org/10.3390/fi14060187, https://www.mdpi.com/1999-5903/14/6/187
5. Carchiolo, V., Malgeri, M.: Navigating the AI timeline: From 1995 to today. In: Proceedings of the 13th International Conference on Data Science, Technology and Applications - Volume 1: DATA. pp. 577–584. INSTICC, SciTePress (2024). https://doi.org/10.5220/0012856700003756
6. Chamola, V., Hassija, V., Gupta, V., Guizani, M.: A comprehensive review of the covid-19 pandemic and the role of iot, drones, ai, blockchain, and 5g in managing its impact. IEEE Access **8**, 90225 - 90265 (2020). https://doi.org/10.1109/ACCESS.2020.2992341
7. Chowdhury, M.Z., Shahjalal, M., Ahmed, S., Jang, Y.M.: 6g wireless communication systems: Applications, requirements, technologies, challenges, and research directions. IEEE Open Journal of the Communications Society **1**, 957 - 975 (2020). https://doi.org/10.1109/OJCOMS.2020.3010270
8. Conoscenti, M., Vetrò, A., De Martin, J.C.: Blockchain for the internet of things: A systematic literature review. In: 2016 IEEE/ACS 13th International Conference of Computer Systems and Applications (AICCSA). pp. 1–6 (2016). https://doi.org/10.1109/AICCSA.2016.7945805
9. Dutta, P., Choi, T.M., Somani, S., Butala, R.: Blockchain technology in supply chain operations: Applications, challenges and research opportunities. Transportation Research Part E: Logistics and Transportation Review **142**, 102067 (2020). https://doi.org/10.1016/j.tre.2020.102067, https://www.sciencedirect.com/science/article/pii/S1366554520307183
10. Dwivedi, Y.K., Hughes, D.L., Coombs, C., Constantiou, I., Duan, Y., Edwards, J.S., Gupta, B., Lal, B., Misra, S., Prashant, P., Raman, R., Rana, N.P., Sharma, S.K., Upadhyay, N.: Impact of covid-19 pandemic on information management research and practice: Transforming education, work and life. International Journal of Information Management **55**, 102211 (2020). https://doi.org/10.1016/j.ijinfomgt.2020.102211, https://www.sciencedirect.com/science/article/pii/S026840122031286X

11. Elsevier B.V.: Elsevier Developer - Academic Research. https://dev.elsevier.com/api_service_agreement.html, accessed: July, 2021
12. Elsevier B.V.: Elsevier Developer - API Service Agreement. https://dev.elsevier.com/academic_research_scopus.html, accessed: Juy, 2021
13. Galvez, J.F., Mejuto, J., Simal-Gandara, J.: Future challenges on the use of blockchain for food traceability analysis. TrAC - Trends in Analytical Chemistry **107**, 222 - 232 (2018). https://doi.org/10.1016/j.trac.2018.08.011
14. Gordon, W.J., Catalini, C.: Blockchain technology for healthcare: Facilitating the transition to patient-driven interoperability. Computational and Structural Biotechnology Journal **16**, 224 - 230 (2018). https://doi.org/10.1016/j.csbj.2018.06.003
15. Griggs, K.N., Ossipova, O., Kohlios, C.P., Baccarini, A.N., Howson, E.A., Hayajneh, T.: Healthcare blockchain system using smart contracts for secure automated remote patient monitoring. Journal of Medical Systems **42**(7) (2018). https://doi.org/10.1007/s10916-018-0982-x
16. Hassija, V., Chamola, V., Saxena, V., Jain, D., Goyal, P., Sikdar, B.: A survey on iot security: Application areas, security threats, and solution architectures. IEEE Access **7**, 82721 - 82743 (2019). https://doi.org/10.1109/ACCESS.2019.2924045
17. Javaid, M., Haleem, A., Singh, R.P., Suman, R., Khan, S.: A review of blockchain technology applications for financial services. BenchCouncil Transactions on Benchmarks, Standards and Evaluations **2**(3), 100073 (2022). https://doi.org/10.1016/j.tbench.2022.100073, https://www.sciencedirect.com/science/article/pii/S2772485922000606
18. Jiang, W., Han, B., Habibi, M.A., Schotten, H.D.: The road towards 6g: A comprehensive survey. IEEE Open Journal of the Communications Society **2**, 334–366 (2021). https://doi.org/10.1109/OJCOMS.2021.3057679
19. Kang, J., Xiong, Z., Niyato, D., Xie, S., Zhang, J.: Incentive mechanism for reliable federated learning: A joint optimization approach to combining reputation and contract theory. IEEE Internet of Things Journal **6**(6), 10700–10714 (2019). https://doi.org/10.1109/JIOT.2019.2940820
20. Kiayias, A., Russell, A., David, B., Oliynykov, R.: Ouroboros: A provably secure proof-of-stake blockchain protocol. Lecture Notes in Computer Science **10401 LNCS**, 357 - 388 (2017). https://doi.org/10.1007/978-3-319-63688-7_12
21. Klerkx, L., Jakku, E., Labarthe, P.: A review of social science on digital agriculture, smart farming and agriculture 4.0: New contributions and a future research agenda. NJAS - Wageningen Journal of Life Sciences **90-91**, 100315 (2019).https://doi.org/10.1016/j.njas.2019.100315, https://www.sciencedirect.com/science/article/pii/S1573521419301769
22. Newman, M.E.J.: Coauthorship networks and patterns of scientific collaboration. Proceedings of the National Academy of Sciences **101**(suppl_1), 5200–5205 (2004). https://doi.org/10.1073/pnas.0307545100, https://www.pnas.org/doi/abs/10.1073/pnas.0307545100
23. Qadri, Y.A., Nauman, A., Zikria, Y.B., Vasilakos, A.V., Kim, S.W.: The future of healthcare internet of things: A survey of emerging technologies. IEEE Communications Surveys & Tutorials **22**(2), 1121–1167 (2020). https://doi.org/10.1109/COMST.2020.2973314
24. Qadri, Y.A., Nauman, A., Zikria, Y.B., Vasilakos, A.V., Kim, S.W.: The future of healthcare internet of things: A survey of emerging technologies. IEEE Communications Surveys and Tutorials **22**(2), 1121 - 1167 (2020). https://doi.org/10.1109/COMST.2020.2973314, cited by: 528
25. Taylor, P.J., Dargahi, T., Dehghantanha, A., Parizi, R.M., Choo, K.K.R.: A systematic literature review of blockchain cyber security. Digital Communications and Networks **6**(2), 147–156 (2020). https://doi.org/10.1016/j.dcan.2019.01.005, https://www.sciencedirect.com/science/article/pii/S2352864818301536
26. Ting, D.S.W., Carin, L., Dzau, V., Wong, T.Y.: Digital technology and covid-19. Nature Medicine **26**(4), 459–461 (2020). https://doi.org/10.1038/s41591-020-0824-5
27. Warner, K.S., Wäger, M.: Building dynamic capabilities for digital transformation: An ongoing process of strategic renewal. Long Range Planning **52**(3), 326–349 (2019). https://doi.org/10.1016/j.lrp.2018.12.001, https://www.sciencedirect.com/science/article/pii/S0024630117303710

28. Xu, M., Chen, X., Kou, G.: A systematic review of blockchain. Financial Innovation **5**(1), 27 (2019). https://doi.org/10.1186/s40854-019-0147-z
29. Zhang, P., White, J., Schmidt, D.C., Lenz, G., Rosenbloom, S.T.: Fhirchain: Applying blockchain to securely and scalably share clinical data. Computational and Structural Biotechnology Journal **16**, 267 - 278 (2018). https://doi.org/10.1016/j.csbj.2018.07.004
30. Zhang, Z., Xiao, Y., Ma, Z., Xiao, M., Ding, Z., Lei, X., Karagiannidis, G.K., Fan, P.: 6g wireless networks: Vision, requirements, architecture, and key technologies. IEEE Vehicular Technology Magazine **14**(3), 28–41 (2019). https://doi.org/10.1109/MVT.2019.2921208
31. Zhu, G., He, D., An, H., Luo, M., Peng, C.: The governance technology for blockchain systems: a survey. Frontiers of Computer Science **18**(2), 182813 (2023). https://doi.org/10.1007/s11704-023-3113-x, 10.1007/s11704-023-3113-x

Large Language Models (LLMs)

Comparative Study on Automatic Extraction of Genealogical Data from Ancient Text by Using Large Language Models and Open AI

Violet Chella Christy Alex Xavier, Md Mahbub Ul Alam, and Doina Logofătu

Abstract Genealogy extraction from ancient texts is a challenging but significant task for genealogical research due to its historical data. To explore each family history, one has to read large amounts of text to find the depth of the relationships between characters. Recent advancements in Natural Language Processing (NLP), especially for information extraction and visualization, have become significantly more accurate, convenient, and interactive due to Artificial Intelligence (AI) integration, enabling them to handle complex language tasks with ever-increasing accuracy through continuous learning from data. This paper first explores the application of modern information extraction (IE) for automated genealogy extraction from ancient texts by incorporating the advanced large language modeling capabilities through OpenAI's GPT models that are integrated into the LangChain framework. This approach not only identifies genealogical information but also interprets and contextualizes relationships using advanced natural language understanding capabilities. The LangChain framework facilitates a seamless connection between OpenAI's GPT (generative pre-trained transformer) models and the Neo4j graph database, enabling the construction of dynamic and context-aware family trees. Then this paper compares the strengths and limitations of the automatic extraction of genealogical data from unstructured ancient texts using GPT models.

Keywords Artificial intelligence · Natural language processing · Information extraction · OpenAI · GPT · Neo4j · LangChain · Python

V. C. C. A. Xavier (✉) · M. M. Ul Alam · D. Logofătu
Frankfurt University of Applied Sciences, Frankfurt am Main, Germany
e-mail: xavier@stud.fra-uas.de

M. M. Ul Alam
e-mail: md.ulalam@stud.fra-uas.de

D. Logofătu
e-mail: logofatu@fb2.fra-uas.de

N. Polatidis et al. (eds.), *Intelligent Distributed Computing XVII*, Studies in Computational Intelligence 1203, https://doi.org/10.1007/978-3-031-87639-4_14

1 Introduction

Natural Language Processing (NLP) has transformed the extraction of information from unstructured texts. Extracting genealogical data from ancient texts, like the Bible, is particularly challenging due to linguistic complexity, varied translations, and non-standard organization. Traditionally, this task is labor-intensive, but recent advancements in large language models (LLMs), such as OpenAI's GPT-3.5 turbo and GPT-4, provide more efficient solutions [8, 9].

This study explores the use of LLMs within the LangChain framework, integrated with the Neo4j graph database, to automatically extract and visualize genealogical relationships from ancient texts. By applying modern NLP techniques, we aim to simplify the process of generating dynamic family trees from historical records [12].

2 Problem Description

Natural Language Processing (NLP) has significantly advanced the understanding and manipulation of human language, revolutionizing fields such as information extraction and human-computer interaction. Extracting family lineage from ancient texts, like the Bible, traditionally involves manual efforts, making it a complex and time-consuming process. This challenge is exacerbated by the unstructured nature of ancient texts, their linguistic nuances, specific terminologies, varied narrative styles, and multiple translations over time [11].

This study addresses these challenges by leveraging state-of-the-art NLP techniques, particularly transformers and large language models (LLMs), to automate the extraction of genealogical data. Specifically, we analyze the application of OpenAI's GPT-3.5 turbo and GPT-4 models within the LangChain framework, integrated with the Neo4j graph database, to reconstruct and visualize family trees from the Bible [1]. This paper analyzes the potential of automatically extracting genetic data from unstructured text and the impact of using LLM in information extraction (Fig. 1).

By exploring these techniques, the study aims to provide insights into the practical application of advanced NLP tools for extracting and contextualizing genetic information from unstructured ancient texts.

3 Literature Review

This paper focuses on the language of the text and the familial relationships between the characters specified in the text. The previous research works that have impacted information extraction with traditional and modern approaches are reviewed here.

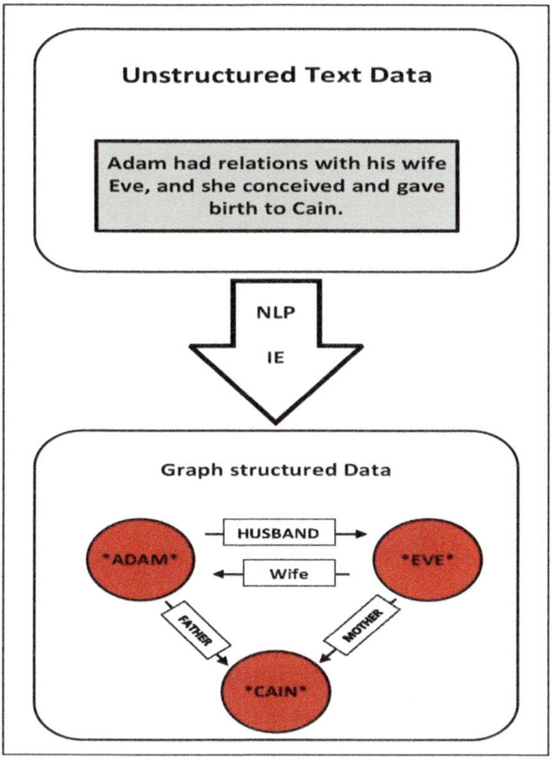

Fig. 1 Automatic extraction of ancient unstructured data to graph structured data using natural language processing with modern information extraction techniques

There has been significant interest in information extraction in the domains of digitized medical, legal, forensic, and publications [15, 16], which are similar to this paper. Most of them use a traditional pipeline to extract information. A few of the related works applied in these fields are discussed below.

Traditional IE Approaches

Medical Informatics: Rybinski et al. (2021) implemented rule-based methods and coreference resolution techniques using BioBERT and Clinical-BERT models to extract family member information from clinical texts. Their work demonstrated the effectiveness of specialized models in handling domain-specific language and complex relationships within medical records [14].

Digital Forensics: In digital forensics, traditional NLP pipelines have been employed to analyze large datasets. For instance, Rodrigues et al. (2022) developed an NLP forensic pipeline using tools like spaCy [3] for named entity recognition and models such as ELECTRA [4], OpenNRE [6] and RoBERTa [5] for relation extraction. Their approach involved constructing directed graphs with Neo4j [7] to visualize and investigate relationships within forensic data [2].

Modern IE Approaches with LLMs

Narrative Entity Extraction: Recent studies have evaluated the performance of large language models (LLMs) like GPT-3 and GPT-3.5 for narrative entity extraction. Sousa et al. (2023) explored the capabilities of these models through prompt-based learning [8], finding that they performed well in generating accurate representations of complex narratives from text.

Advancements in GPT Models: Rahaman et al. (2023) compared GPT-3 and GPT-4, highlighting significant improvements in model capabilities and performance. Their study noted that GPT-4's advanced language understanding and generation abilities made it more effective for information extraction tasks [9].

Few-Shot Learning: Wadhwa et al. (2023) demonstrated that GPT-3 outperforms other models like Flan T5 in relation extraction tasks using few-shot learning techniques [10]. This approach allows the model to learn from a small number of examples, enhancing its ability to extract relationships accurately from text.

Integration with Graph Databases

Neo4j and Graph Representation: Graph databases like Neo4j have become increasingly popular for representing complex relationships in extracted data. Neo4j's flexibility and powerful query capabilities make it an ideal choice for visualizing genetic data [7]. The integration of LLMs with Neo4j, as facilitated by the LangChain framework, enables dynamic and context-aware construction of family trees from unstructured texts.

LangChain Framework: The LangChain framework provides a robust infrastructure for integrating LLMs with various data processing and storage tools. Its modular design allows for seamless connection between language models and graph databases, enhancing the automation and accuracy of information extraction processes [12, 13].

The reviewed literature underscores the rapid progress in IE techniques and the potential of LLMs to enhance the extraction and interpretation of genealogical data from ancient texts. Traditional methods, while effective in specific domains, often fall short in handling the complexity and scale of unstructured historical texts. Modern LLMs, integrated with frameworks like LangChain and databases like Neo4j, offer promising solutions for these challenges. By leveraging these advancements, our study aims to improve the accuracy and efficiency of genealogical data extraction using state-of-the-art NLP tools.

4 Methodology

This approach employs LangChain [19] for data integration and retrieval of LLM with the Neo4j graph database [20]. The workflow starts with requesting the unstructured text data from the web and passes it into the integration framework, LangChain. The methodology consists of two primary approaches: LLM processing with summarization and direct LLM processing. The data is then tokenized and split into smaller chunks of documents for better processing using LangChain retrieval methods. The dataset poses challenges due to the large volume of data, linguistic issues from multiple translations, and the variety of familial relationships that need to be identified and contextualized.

The LLM processing with summarization is that the text is divided into manageable chunks to facilitate processing. This is done using the MapReduce approach within the LangChain framework, which efficiently handles large volumes of data by parallelizing the summarization task. Each chunk is summarized using the GPT-3.5 turbo model. The summarization aims to condense the information while retaining the key genealogical relationships.

Next, the summarized chunks are further processed to identify genealogical entities and relationships. LangChain employs predefined prompts and templates to extract structured data, creating nodes (entities) and edges (relationships) that represent the genealogical information. The extracted data is formatted into a structured schema suitable for graph databases and stored in Neo4j.

The summarization process involves generating concise summaries of each document chunk, followed by merging these summaries into a cohesive whole. This step ensures that critical information is not lost during the summarization. The summarized data is then split into smaller chunks for processing by the LLMs, which generate the structured genetic data.

During direct LLM processing, the summarization may be a reduction in the cost, but it might extract the most important entities alone, leaving others. To identify the difference between models, the tokenized chunks of documents are given directly to the LLM model for creating nodes and edges representing entities within the text. This approach captures the most critical entities, potentially revealing additional information compared to the summarized data.

Both approaches store the data in the Neo4j graph database, providing visual insights into the genealogical data described in Fig. 2.

The Large Language Model (LLM) is trained on massive amounts of text data. They rely on complex neural network architectures like transformers [17] for learning and processing language. With careful development and responsible use, these models have the potential to revolutionize how humans can interact with language and information. It is getting popular with generative AI and low-code/no-code. This work analyzes the capabilities of the GPT 3.5 Turbo and GPT 4 models as they specialize in NLP tasks. The LLMs can remember the structures and patterns of different languages and can generate content based on prompts.

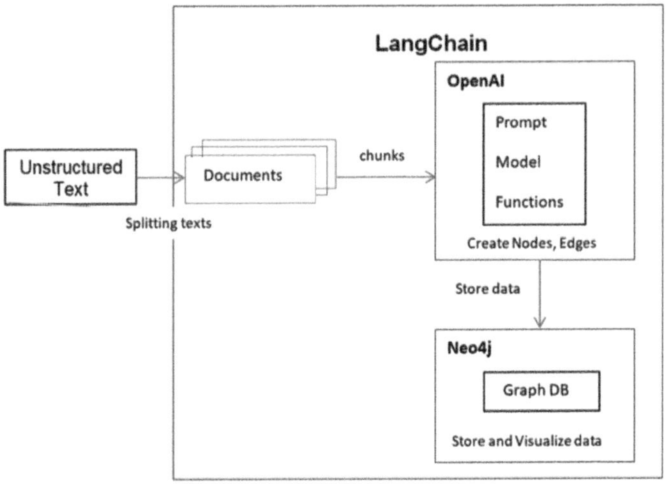

Fig. 2 The process of LangChain integration with LLM and Neo4j for extracting the unstructured ancient text and forming graph-structured data

Prompt engineering focuses on crafting instructions and information for AI models to shape their responses in desired ways. It aligns directly with NLP's goal of enabling computers to work with and understand human language. The model can perform specific NLP tasks like text generation, translation, or summarization by carefully constructing prompts. Effective prompts can unlock capabilities beyond the model's initial training data, demonstrating AI development's creative and strategic aspects. The practical guidelines that helped this work for writing prompts [18] are using delimiters to distinguish the data from the prompt and asking for structured output such as JSON (JavaScript Object Notation), XML (eXtensible Markup Language), and HTML (HyperText Markup Language). It can include information to modify the tone of the output, setting conditions to the model, and asking if those conditions are verified by providing successful examples similar to the tasks and then asking to perform the task. It performs better when the steps required to achieve a task are specified. Finally, iterate and refine the prompts.

Graph databases are used to represent complex relationships in data. They focus on connections between entities, allowing for efficient analysis and visualization of those relationships. The data is stored in a schema-flexible way, allowing them to grow dynamically. It has a graph query language, Cypher, for extracting and visualizing complex relationships. The graph database is chosen according to the performance, requirements, data size, and cost of the specific work. This work utilizes the Neo4j [20] for expressing the genealogically interconnected family tree. It is a user-friendly graph database management system with little free native graph storage.

For the LLM to integrate with the above tools and services, LangChain a seamless integration framework is used. It can build complex workflows according to specific needs. This framework has built-in libraries and templates for various services and components for external integration. This work needs an integration framework for combining prompts, models, and graphs from different providers. It has modules for retrieving data from various sources and agents, providing multiple functions and chains for document processing. Here, LLM prompts, Document Loader from the retrieval module, OpenAI functions from the agent module, MapReduce summarization from the chain module, and Neo4j graph from the graph community services are utilized.

OpenAI's GPT language models, which are specifically designed for NLP, were chosen. The pricing plans [21] differ with each model's input and output tokens. Therefore, this work tried to find the accuracy in two ways. One is to summarize the input data and then fetch the resulting output. Another way is to give the entire input data and fetch the result.

4.1 Data Source

Here, extracting genealogical data from ancient texts is the main concern. To have unstructured ancient text data, the Berean Translation Bible [11], which has free downloads and free licensing for accessing the contents, is used.

The following are some of the challenges this work must consider while extracting. They are,

• Large amount of data:
The Bible consists of numerous books and chapters. It has a wide array of names, including short names, different names for the same characters, and the same name given for different characters. These may result in complexity and inaccuracy.

• Linguistic Issues:
As the Bible is one of the ancient books, many versions of translations in different languages exist. The challenge comes with the understanding of linguistic variations in narrations and words.

• Variety of relationships:
As discussed, many names can be identified. However, this study aims to extract and form a family tree for each name. To fetch the relationships between each name is more complex because of indirect poetic lineage narrations in the context.

4.2 LLM Processing with Summarization

Here two steps take place. First summarize, then extract graph structure.

(i) Summarization: The documents are processed through the summarization process using the MapReduce approach. It involves summarizing each document chunk using the same LLM prompt template and forming summarized smaller chunks. These are then merged and again summarized into one cohesive summary,

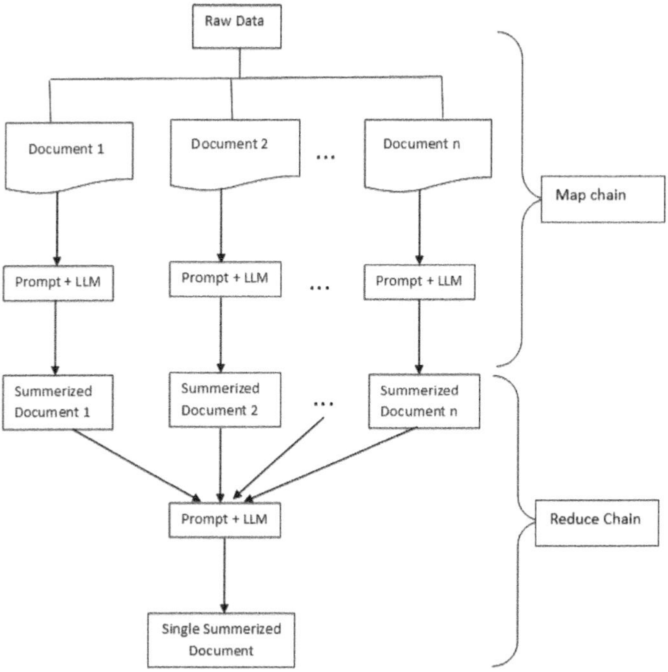

Fig. 3 The process of MapReduce Summarization technique using LangChain framework which integrates the LLM with prompts

as explained in the Fig. 3. Using the MapReduce method [12], the model can summarize extensive documents while chaining the context limitations of the Stuffing method through parallel processing. However, the MapReduce requires multiple calls to the model and potentially losing context between pages. The Chain of Responsibility design pattern is followed in the processing steps through the chain of handlers according to the need in MapReduceDocumentsChain, ReduceDocumentsChain and StuffDocumentsChain.

The MapReduceDocumentsChain takes the document input to perform Map and Reduce operations sequentially. The ReduceDocumentsChain reduces the document size to the fixed chunk size when it exceeds. The StuffDocumentsChain wraps another chain and adds functionality to handle splitting and combining documents. MapChain follows a template design pattern. Whenever there is a need to map either the split document or the summarized document, it uses the template to enable the prompt and model. Now the summarized document for the large data is stored.

(ii) Data extraction process: The LangChain uses the OpenAI model and functions to extract the data. It sequentially creates the nodes and edges for the graph. First, the summarized document is again split and sent as chunks of documents to create structured output. Then, the prompt is given by explaining the role of the request and the steps to be followed for each problem. This extracts the entities and their relationships as given in the LangChain template documentation [13]. Next, the extracted data creates the nodes and edges using the graph document, which has the template for filling the values for nodes and edges. The datatype to be stored is also validated using the pydantic module. The nodes and their relationships are added according to the extraction. The graph is stored in JSON format with the property values. The graph is stored directly and viewed in the Neo4j Aura database. In this context, two prompt templates are created for the initial summarization step and the final consolidation step for extracting data.

4.3 *Direct LLM Processing*

In the first process, the gpt3.5turbo model is used to summarize the large data and then form a graph structure. Here, the attempt to use the GPT4 model is processed. The sequence diagram for the direct LLM processing is given in Fig. 4. The raw data is given as input into the LangChain framework. It then generates documents using document splitters and uses prompts for creating graphs. The graph schema defines the structure of the graph that will be used to store the data. The nodes and edges are created according to the attributes needed for the Neo4j graph database. Finally, the nodes and edges are stored in the Neo4j graph database, where they can be visualized and queried.

5 Experimental Results and Challenges

This paper uses the GPT models, which learn from large amounts of pre-trained data. Here, the large amount of data is summarized and then used to form a graphical structure in Neo4j.

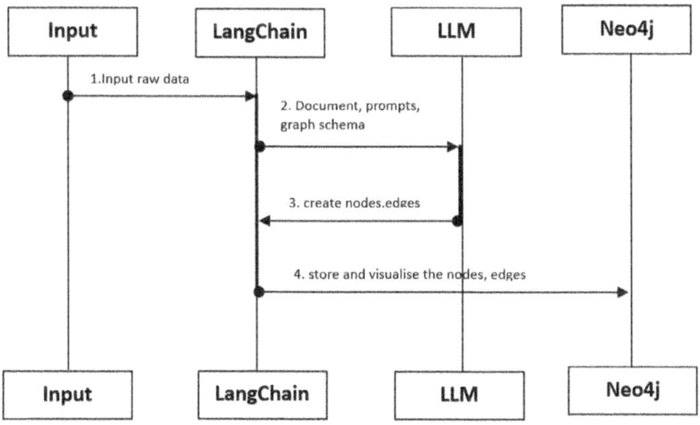

Fig. 4 Sequence diagram for direct LLM processing

5.1 Experimental Results

The first process tried to summarize a large amount of data and then form a graphical structure using Neo4j. The LangChain framework integrates different modules needed for this specific outcome. For summarization, gpt3.5-turbo was used. And the outcome was cost-effective. Another method is used by directly giving the prompt without the summarization process for the entire large amount of data using the gpt4 model, which is faster, more accurate and more expensive than the gpt3.5-turbo model.

The Fig. 5 gives the visual appearance of the created nodes and edges. Here, Nodes are Person names, and Relationships are specified relation-mentions. The family relationships focused in this work are Father, Mother, Husband, Wife, Concubine, Son, Daughter, Child, Brother, Sister, Grandson, Granddaughter, Grandchild, Grandfather, Grandmother, Father-in-law, Mother-in-law, and Uncle.

However, a few other types of partnerships were also mentioned, such as FRIEND and DISCIPLE. This is one of the key distinctions between traditional NLP models and generative models in extracting data. Traditional models excel at targeted extraction, trained specifically to identify and extract predefined data points. This ensures high accuracy for the intended task but limits their flexibility. Generative models, on the other hand, leverage extensive text embeddings to potentially extract information beyond the explicitly trained data. This offers exciting possibilities for uncovering dynamic insights. However, the automatic nature of this extraction necessitates careful evaluation to ensure the generated information is reliable. While generative models hold immense potential, their dynamic outputs require a robust validation process.

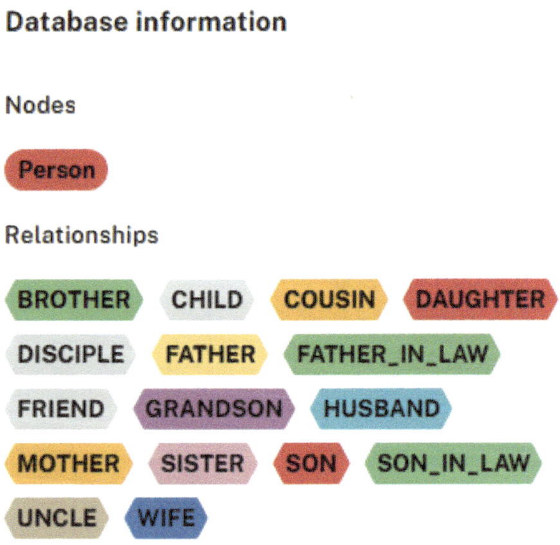

Fig. 5 Dynamic graph generation using GPT models. The nodes represent characters, and relationships depict possible familial connections extracted from the given data

Fig. 6 Attributes for a Person Node in Neo4j containing id and a key-Value pair

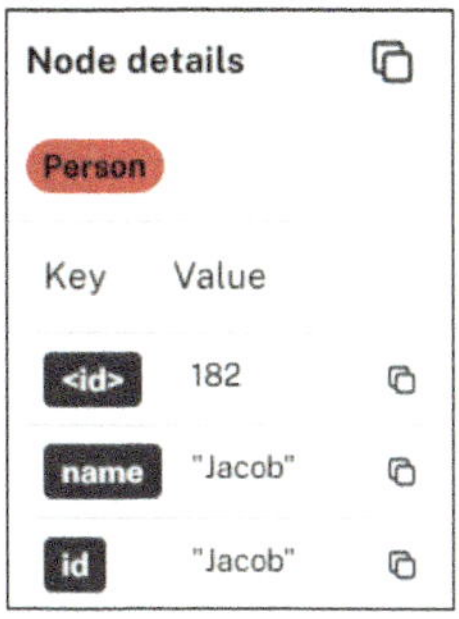

Each node is assigned with the graph field attributes, which is the Key-value pair as in Fig. 6, along with the validation during the node creation process.

The results are compared between the summarization process and the direct process of LLM by extracting a few important characters in the Bible.

Effect of different characters with the same name: In this family tree (Figs. 7 and 8), the names 'Enoch' and 'Lamech' come in subnodes of the interconnected family tree of 'Cain' and 'Seth' but they are different characters with the same name. This is achieved accurately with the gpt4 Fig. 8. The summarization process using gpt3.5-turbo splits and summarizes the document many times and cannot find the difference between the characters. It confuses with similar names, which are actually from a different family tree.

Effect of the same character with a different name: In this family tree (Figs. 9 and 10), the characters 'Abraham' and 'Sarah' changed their names to 'Abram' and 'Sarai' in later history. The gpt4 model has identified this similarity of names and also gave a detailed family tree but still, gpt3.5-turbo achieved the tree with different names for the same character.

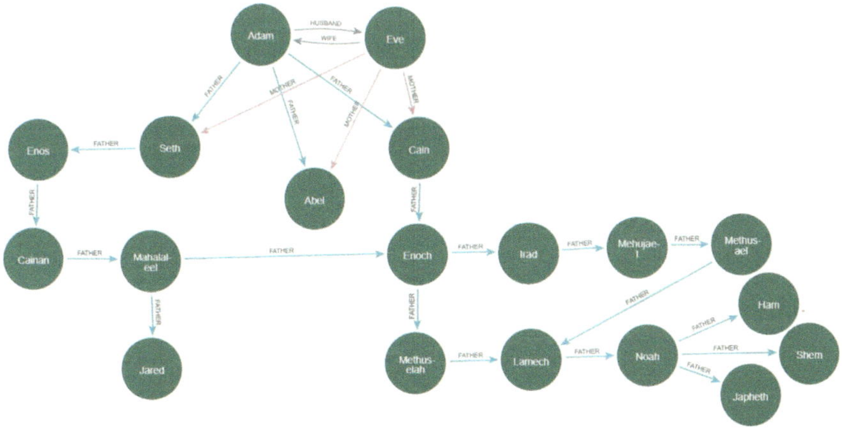

Fig. 7 Adam–Family Tree using summarization process with gpt3.5-turbo model

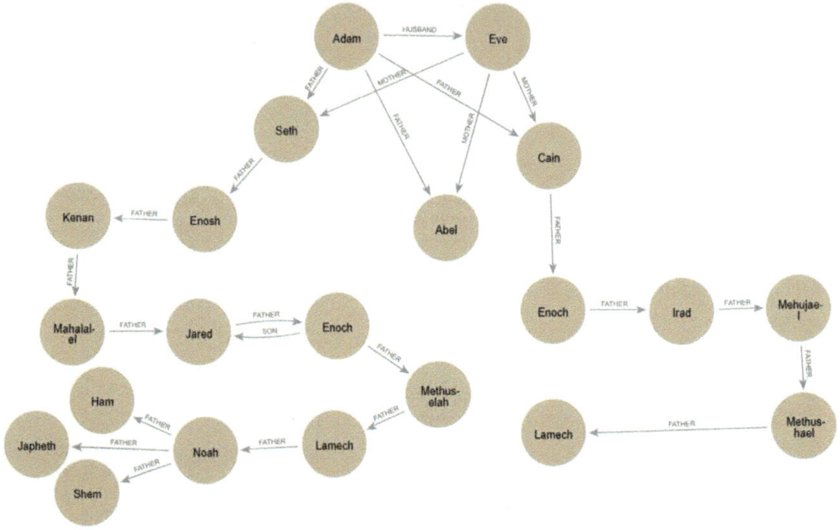

Fig. 8 Adam–Family Tree using gpt4 model without the summarization process

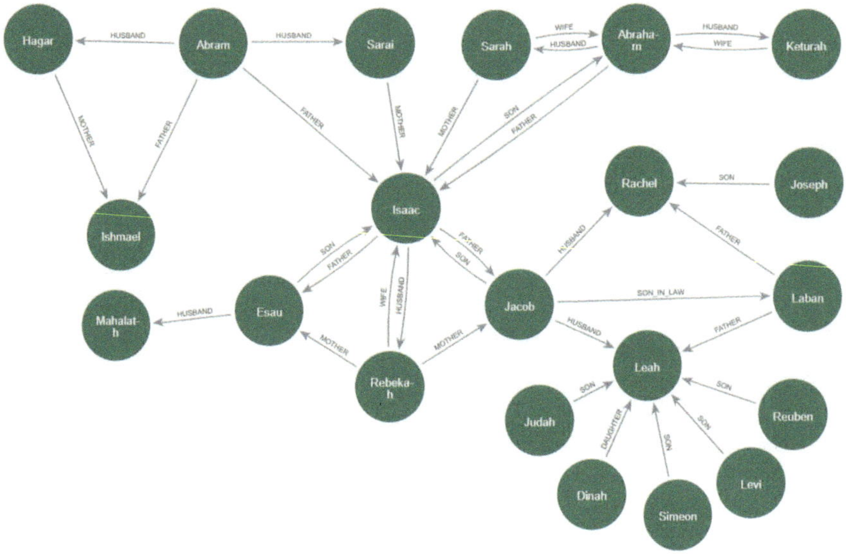

Fig. 9 Abraham–Family Tree using summarization process with gpt3.5-turbo

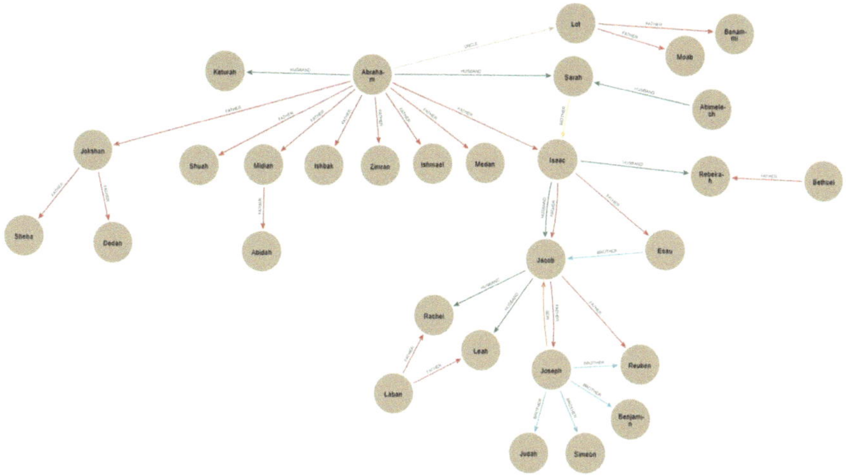

Fig. 10 Abraham–Family Tree using gpt4 model without the summarization processAbraham–Family Tree using gpt4 model without the summarization process

5.2 Challenges

The challenge of managing extensive data is met with advancements in processing power and storage solutions for accurate results. Although the output of the LLM is faster and more accessible, the accuracy increases along with the cost. OpenAI's pricing plan for processing a large amount of data with their GPT models must be carefully estimated for the specific use case. The final prompt is chosen after trying different prompts to get the specific result because the GPT models learn from large amounts of data and give a result that matches the specific mentioned in the prompt. Extracting large amounts of data cannot be ensured with specific outcomes. The process is more convenient than the traditional ML process, but the accuracy of the output needs manual proofing. The data privacy and security are limited, which raises concerns while extracting with free cloud services provided by Neo4j and lack of transparency with GPT models. Given the historical nature of the extracted data, priority must be given to the meticulous approach of content generation, ensuring factual accuracy and avoiding any potential for misrepresentation. The GPT models can be a powerful tool for automatic data extraction, and their outputs can appear dynamic due to their ability to identify implicit patterns. Despite the limitations, the advancement in NLP has exponential growth with memory, storage and security.

6 Conclusion and Future Work

This study demonstrates the potential of combining advanced large language models (LLMs) with the LangChain framework and the Neo4j graph database to automate the extraction of genealogical data from ancient texts. By leveraging OpenAI's GPT-3.5 turbo and GPT-4 models, significant improvements were achieved in the accuracy and efficiency of information extraction compared to traditional NLP methods. The two approaches explored-summarization followed by extraction, and direct processing-offered different trade-offs in terms of cost, accuracy, and computational efficiency. The summarization approach, using GPT-3.5 turbo, was more cost-effective but less accurate, while the direct processing approach, using GPT-4, provided higher accuracy at a greater cost. These findings highlight the importance of selecting the appropriate method based on specific requirements and resource constraints.

Overall, this study contributes to the fields of natural language processing, historical research, and genealogy by providing a novel methodology for extracting and visualizing complex genealogical relationships from unstructured ancient texts. The integration of LLMs with LangChain and Neo4j offers a powerful toolset for handling large volumes of text data and generating detailed, context-aware family trees.

In the future, this work aims to refine the extraction processes by incorporating cutting-edge techniques for similarity analysis and path-finding to enhance the accuracy of character identification and relationship mapping. We will investigate the use of additional LLMs and transformer models to improve the robustness and reliability of the information extraction process.

Developing robust quantitative metrics to evaluate the accuracy and efficiency of the extraction process is essential. This includes creating benchmarks for precision, recall, and F1 scores of different models and approaches. Solutions must be established in the future to have robust evaluation metrics. It might be possible by linking the entities with real-time data using the Large Multimodel Model (LMM), which works with generative AI and various forms of data such as images, videos, audio, and other related text data. The latest iterative and fast-paced advancement of LLM holds the potential for unforeseen developments in the future.

References

1. List of biblical characters, Online:https://www.wikidata.org/wiki/Wikidata:Lists/List_of_biblical_characters, last accessed 2024/03/23
2. F. B. Rodrigues, W. F. Giozza, R. de Oliveira Albuquerque and L. J. García Villalba, "Natural Language Processing Applied to Forensics Information Extraction With Transformers and Graph Visualization," in IEEE Transactions on Computational Social Systems", 2022, https://doi.org/10.1109/tcss.2022.3159677
3. Neuralcoref, Online:https://github.com/huggingface/neuralcoref, last accessed 2024/03/23.

4. Clark, Kevin, Minh-Thang Luong, Quoc V. Le, and Christopher D. Manning. "ELECTRA: Pre-training text encoders as discriminators rather than generators", 2020, https://doi.org/10.48550/arXiv.2003.10555.
5. Yinhan Liu, Myle Ott, Naman Goyal, Jingfei Du, Mandar Joshi, Danqi Chen, Omer Levy, Mike Lewis, Luke Zettlemoyer and Veselin Stoyanov, "RoBERTa: A Robustly optimised BERT pretraining Approach", 2019, https://doi.org/10.48550/arXiv.1907.11692
6. OpenNRE framework, Online: https://github.com/thunlp/OpenNRE, last accessed 2024/03/23.
7. Neo4j - a graph database management system, Online: https://neo4j.com/, last accessed 2024/03/23.
8. H. Sousa, N. Guimaraes, A. Jorge and R. Campos, "GPT Struct Me: Probing GPT Models on Narrative Entity Extraction," in 2023 IEEE International Conference on Web Intelligence and Intelligent Agent Technology (WI-IAT), Venice, Italy, 2023 pp. 383-387, https://doi.org/10.1109/WI-IAT59888.2023.00063
9. Rahaman, Md. Saidur, Ahsan, M. M. Tahmid, Anjum, Nishath, Terano, Harold Jan, Rahman, Md., " From ChatGPT-3 to GPT-4: A Significant Advancement in AI-Driven NLP Tools", Journal of Engineering and Emerging Technologies. 1. 50-60, 2023, https://doi.org/10.52631/jeet.v1i1.188.
10. Wadhwa, Somin, Silvio Amir and Byron C. Wallace. "Revisiting Relation Extraction in the Era of Large Language Models." Proceedings of the conference. Association for Computational Linguistics. Meeting 2023: 15566-15589. https://doi.org/10.48550/arXiv.2305.05003
11. Data used, Online: https://bereanbible.com/bsb.txt, last accessed 2024/03/23.
12. LangChain Summarization, Online: https://python.langchain.com/docs/use_cases/summarization, last accessed 2024/03/23.
13. LangChain chains, Online: https://api.python.langchain.com/en/latest/chains/langchain.chains.ernie_functions.base.create_structured_output_chain.html, last accessed 2024/03/23.
14. Rybinski M, Dai X, Singh S, Karimi S, Nguyen, "Extracting family history information from electronic health records: Natural language processing analysis", JMIR med inform 2021, https://doi.org/10.2196/30153
15. Markus krug, "Techniques for the automatic extraction of character networks in German Historic Novels", 2020, https://doi.org/10.25972/opus-20918.
16. Timothy Sui-Tim Chu, "Genealogy Extraction and Tree Generation from Free Form Text", 2017, https://doi.org/10.15368/theses.2017.112
17. Mandar Joshi, Danqi Chen, Yinhan Liu, Daniel S. Weld, Luke Zettlemoyer, and Omer Levy. "SpanBERT: Improving Pre-training by Representing and Predicting Spans", Transactions of the Association for Computational Linguistics, 8:64-77, 2020,https://doi.org/10.48550/arXiv.1907.10529
18. Prompt Engineering, Online: https://platform.openai.com/docs/guides/prompt-engineering, last accessed 2024/03/23.
19. LangChain Framework, Online: https://python.langchain.com/docs/get_started/introduction, last accessed 2024/03/23.
20. Neo4j Graph Database, Online: https://neo4j.com/, last accessed 2024/03/23.
21. Openai pricing, Online: https://openai.com/pricing, last accessed 2024/03/23.
22. Bratanic, Tomaž, "Graph Algorithms for Data Science: With Examples in Neo4j.", Simon and Schuster, 2024.

Planless Agents: BDI Plan Generation Using LLMs

Theresa Schulz and Kai Jander

Abstract Belief-Desire-Intention (BDI) agents offer a unique approach for engineering complex behavior for individual agents in multi-agent systems. A developer can define goals for each agent, specifying the desired outcomes in various contexts and implement plans as the means to reach those goals. The BDI reasoning engine can then proceed to automatically select goals to pursue (goal deliberation) and choose one or more of the provided plans to attempt to achieve them. However, usually plans or at least part of the plans have to be provided by the agent's developer before the system is deployed. In this paper we present an approach for BDI agents to generate their own plans using large language models (LLMs) using solely the available context information such as goal descriptions, available beliefs and their structural information. We show that the approach is viable in principle and explore its reliability as well as discuss further use of LLMs in the context of automating BDI-based agents.

1 Introduction

Although Multi-Agent Systems (MAS) is a well-researched domain within the field of AI, their full potential remains underexplored [5]. Particularly combinations of different AI techniques with multi-agent technology may yield new and interesting results that would be unachievable by either technique alone. Recent developments in machine learning have shown the potential of Large Language Models (LLMs) to generate not just natural language text but also generate executable code in popular programming language and as a results there is renewed interest in agent technology and a growing interest in LLM based autonomous agents. These agents are integrated

T. Schulz (✉) · K. Jander
Brandenburg University of Applied Sciences, Brandenburg an der Havel, Germany
e-mail: schuther@th-brandenburg.de

K. Jander
e-mail: jander@th-brandenburg.de

N. Polatidis et al. (eds.), *Intelligent Distributed Computing XVII*, Studies in Computational Intelligence 1203, https://doi.org/10.1007/978-3-031-87639-4_15

into new frameworks and are specifically designed to generate plans that incorporate user-provided constraints and automate the planning process accordingly. Using LLMs, these agents are able to operationalize their internal goals by generating executable code. However, initial attempts do not always yield good results due to factors such as prompt robustness, hallucination and knowledge boundaries [14, 24].

The BDI architecture allows the modelling of agents with human-like behaviour, but it is time-consuming to design agents that exhibit flexible behavior and is sometimes error-prone, especially in complex scenarios [2], since the agent developer may not be able to accurately predict every scenario the agent may encounter. The use of LLMs such as GPT-3 or GPT-4 (Generative Pre-trained Transformer) could offer a solution by assisting with the complexity of agent design and allowing agents to adapt to situation that were not predicted during the design phase. LLMs have demonstrated the ability to understand and generate natural language as well as more formal languages such as general purpose programming language or even agent-specific languages. Automating the BDI architecture with LLMs combines the flexible approach of code generation using LLMs with the structured reasoning provided by the BDI agent model and could potentially save developers time and resources otherwise spent on detailed action planning, coordination, and programming. Consequently, agent systems could be developed more efficiently with less skill and effort. Agents could adapt more swiftly and effectively to diverse application areas. Such an approach could both make BDI agent development more accessible and enhance the adaptability of the BDI model. Furthermore, it would add explainability to the system using the BDI agent's goals as reasons for an agent's actions, which is a feature currently unavailable to pure LLM-based approaches. Potential applications span from personalized assistance systems to intelligent virtual characters (BOTS) in video games or efficient control systems for autonomous robots [13, 16], but could potentially be much broader once more refined methods are developed. The unexplored combination of BDI architecture and LLMs presents an opportunity for contributions to the advancement of artificial intelligence through transformer models and agent technology.

The structure of this paper is as follows. In Sect. 2, we explore the theoretical underpinnings of BDI agents and LLMs, explaining the basic concepts and the synergy between these paradigms. In Sect. 3, we review the existing literature on planning mechanisms for BDI agents, highlighting different strategies and their limitations. Our novel approach to integrating LLMs with Jadex for dynamic plan generation is detailed in Sect. 4. This includes methods for providing contextual information to the LLM, techniques for effective prompt engineering, and the integration of generated plans into executable code. Sect. 5 demonstrates the application of our approach through a case study, and evaluates the performance and identifies any shortcomings in the implementation. Finally, Sect. 6 provides concluding remarks and discusses potential avenues for future research aimed at further enhancing the capabilities of BDI agents with advanced LLM technologies.

2 BDI Agents and Large Language Models (LLMs)

The BDI architecture, rooted in Bratman's theory of intentions, serves as a well-established approach for modelling agent-based systems and giving them the capacity for human-like task-solving behaviour [25]. Bratman's theory [1] describes how agents develop intentions through the interplay of beliefs, desires, and practical reasoning, leading to the formulation of plans to achieve their goals. Using this theory, [18] constructed a framework that uses BDI architectures to represent agents' mental states, enabling the generation of plans based on intentions and practical reasoning. This framework enables agents to navigate through complex environments using their intentions and cognitive processes to efficiently achieve desired outcomes.

The BDI Architecture delves into how these components are represented, updated, and processed to drive agent actions effectively. It defines the structure and mechanisms through which beliefs, desires, and intentions are managed within an agent, ensuring coherence and consistency in decision-making and action selection. This architecture encompasses a belief base, where the agent's knowledge about the world is stored, a set of pending events or goals representing desires, a plan library containing rules to handle events based on contextual conditions and an intention base which holds partially instantiated plans that the agent is committed to. By meticulously orchestrating the interaction between these components, BDI architectures enable agents to navigate complex environments, adapt to changing circumstances and pursue their objectives with purpose and efficiency. A multitude of implemented BDI systems and formally specified languages exist, reflecting the ongoing exploration and refinement of this approach to agent-oriented programming [20].

Despite its effectiveness, the BDI architecture faces challenges in efficient implementation, especially when it comes to dynamic plan generation, coordination of complex tasks and unpredictable environments [21]. Manual development of rules and action sequences for BDI agents can be time-consuming and often requires extensive domain expertise. According to Bordini, integrating AI techniques within the BDI architecture offers significant advantages. The alignment of AI with BDI enhances the system's comprehensibility and verifiability, while also improving context sensitivity and robustness. It is the responsibility of the agent platform developer, to ensure that AI implementations integrate seamlessly with the BDI architecture. The integration is made possible by using pluggable components, which allow for the inclusion of capabilities like reasoning, planning, and scheduling into the agent's functionality [2].

Language models with attention layers, such as LLMs, have revolutionating NLP and become essential in numerous applications and have demonstrated significant potential [22]. In addition to their primary function of language generation, LLMs exhibit capabilities that extend to code generation [9]. Pre-training on extensive text corpora, followed by fine-tuning for specific tasks or domains, empowers LLMs to adapt and specialize their knowledge for diverse applications and contexts. Their large scale, with parameters numbering in the hundreds of billions or trillions, equips them with formidable knowledge representation and inference capabilities, often rivaling or surpassing human performance. Although LLMs may produce outputs that are deemed unreliable or inaccurate, a robust agent development framework can enable a comprehensive evaluation of the generated results for their robustness, reliability and accuracy. This integration guarantees that the generated outputs meet the required standards of accuracy and validity within the specified context [27]. In BDI agent development, integrating LLMs offers a promising solution to challenges associated with manual rule-based programming and plan generation. It also addresses the limitations of single-answer responses and forgetfulness [13]. BDI agents can use LLMs to generate plans by reasoning about goals, beliefs, and intentions expressed in natural language. This automated approach streamlines development and enhances adaptability and responsiveness in dynamic environments. Moreover, integrating LLMs with BDI architecture aligns with the current trend of using AI techniques to enhance the intelligence and autonomy of agent-based systems. This, in turn, promotes the emergence of more sophisticated and human-like behaviours in artificial agents [2].

Concepts for generating autonomous agents using LLMs are currently being developed. There are already innovative examples that illustrate the potential of this technology. One such example is the Generative Agent, a realistic simulation in which 25 agents populate a sandbox world and mimic human behaviour. Users can observe and intervene as the agents plan their day, communicate, form relationships and coordinate activities [16]. Another example is AgentGPT, a browser-based platform that allows users to create and deploy their own autonomous AI agents directly in the browser. Users can specify goals for AgentGPT to understand, develop strategies and deliver results without human intervention [19]. Similarly, there's AutoGPT, a concept that's gaining interest because it can be hosted on your own machine. However, questions remain about its effectiveness and flexibility in real-world decision making [10, 26]. While some of these tools are browser-based and flexible for users, they require prompt engineering, making them less accessible.

All these examples are based on new frameworks and do not use traditional solutions. The integration of Jadex, a framework for building BDI-based agents with LLMs, represents an innovative approach to autonomous agent development. This integration aims to extend the capabilities of Jadex agents through the language processing capabilities of LLMs. Jadex already provides a foundation for creating complex agent systems based on goals, beliefs and intentions. By integrating a LLM, these agents could be empowered to handle tasks requiring advanced text processing and decision making, as found in tools such as AgentGPT and AutoGPT.

3 Related Work

In the BDI architecture, planning plays a central role, serving as the mechanism by which agents navigate their environment to achieve desired goals. In essence, planning can be conceptualised as a search algorithm used to identify optimal sequences of actions that lead to the realisation of pre-defined goals within a given problem space. As such, its output is a structured sequence of actions, designed to meet the challenges posed by the agent's environment and effectively guide it towards the fulfilment of its goals. In scenarios characterised by complexity or unpredictability, pre-defined plans created by developers may prove inadequate to deal with the numerous challenges that emerge. To deal with such situations, the use of automated planners or AI planners within the realm of automated planning and BDI agents becomes imperative. These sophisticated tools use various algorithms and implementations that have evolved over time to effectively generate plans tailored to dynamic and complex environments [6].

BDI systems are constrained primarily by their lack of inherent foresight and planning capability. Planning plays a critical role in avoiding wasted resources, irreversible actions, time inefficiencies and unintended consequences. In conventional BDI setups, actions are initiated in response to shifts in beliefs or goals, with plans added to a queue for execution accordingly. However, there are instances where plans fail to achieve their intended results, raising doubts about the feasibility of the overarching goal.

To address these limitations, recent iterations of BDI systems such as JACK and Jason adopt a trial-and-error approach, testing different plans until the desired goal is achieved. However, this method requires the programmer to explicitly incorporate alternative plans into the system. Furthermore, in traditional BDI setups, the success of a plan depends directly on the achievement of the goal, which can lead to setbacks if plans are not chosen wisely [15].

Bordini et al. explore in detail the integration of AI into different stages of the BDI architecture. Within each cycle, agents programmed within a BDI-based framework engage in decisions about their further actions, including identifying and developing plans to achieve goals. This deliberation requires the application of AI methods, in particular machine learning, to navigate the complexities of decision making. Research efforts have aimed at seamlessly integrating AI techniques into BDI-based agent programming languages, highlighting their central role in refining plan selection and execution processes. While their discourse broadly traces the BDI cycle, encompassing the sense, plan, and act phases, they highlight the nuanced changes and ramifications of AI integration, particularly emphasizing its importance in enhancing the planning phase within the context of their stated goals[2].

The integration in the planning phase aims at dynamic planning by incorporating classical planning methods into agent languages to generate new plans at runtime. There have been efforts to integrate plan generation methods into BDI architectures, such as using a state-based search algorithm working on an agenda of beliefs in Jadex [23], or employing planning formalisms with PDDL in GOAL [12]. Similarly, efforts have been made to use classical first-principles planning to discover plans not included in the agent's plan library [7]. This approach focuses on generating abstract, hybrid plans containing abstract operators. These operators can then be mapped back to BDI goals, allowing the agent to execute the generated plan using its developer-provided BDI plan library.

All these methods are initial steps toward integrating planning methods in BDI architectures. However, they are challenging to implement because they require detailed definitions, adaptation to the system's structures, and significant programming effort, which is time-consuming and resource-intensive.

The authors Herzig et al. recommend a re-evaluation of the relationship between BDI models and planning. They argue that existing decomposition methods are too rigid to adequately represent instrumentality relationships between intentions. This points to the need to integrate more flexible and adaptive approaches into BDI logic in order to better represent the dynamics of intentions [11]. One promising field is the combination of autonomous agents with LLMs in various application areas. The focus is on automating problem solving, integrating with human workflows, and developing frameworks for autonomous agents in conjunction with LLMs [24].

4 LLM-Based Plan Generation

The evolution of planning in BDI systems has seen significant progress over the decades. Early BDI frameworks, as Bordini et al. [3] observed, primarily relied on AI in static, predefined contexts for planning. This approach often led to inflexibility in handling new situations, resulting in inefficient or failed executions. To address this, Bordini and colleagues proposed that agents should plan dynamically in real-time based on their current beliefs, goals and environmental conditions. Our work builds on this notion, advancing BDI systems by enabling dynamic, on-the-fly plan generation without requiring complex planning algorithms, such as those discussed by Silva et al. [8]. While these algorithms offer greater flexibility, they come with significant computational costs. In contrast, static programming or failed executions due to a lack of control mechanisms further highlight the limitations of traditional BDI approaches.

Our approach addresses these limitations by integrating Large Language Models (LLMs) into the BDI architecture. With this system, plans are generated dynamically and executed in real-time, enhancing the agent's flexibility and adaptability to diverse scenarios. While the current implementation focuses on the automation of plan generation, a future direction of our research is for the LLM to generate step-by-step plans incrementally. This would allow the agent to progressively work toward achieving its goals, dynamically adjusting its steps based on the environment and its current state, thus further improving adaptability and robustness.

The standard approach for plans in BDI agents is the implementation of plan or in case of the use of planners, plan fragments as code, which are then executed by the BDI agent when a particular plan is chosen by the agent. This code is usually prepared by humans in advance of BDI execution, supplying the agent with the means to interact with its environment. While, as shown in the previous section, some effort has been made to reduce coding effort by BDI agent developers, ultimately the implementation of such code is still required. The following section will explore how this process could potentially be automated using LLMs, by providing the LLM with the context and situation the agent finds itself in and letting the LLM generate the appropriate code.

The approach was implemented using the Jadex Active Component agent platform. This integration aligns with the core principles of the BDI architecture, as shown in Fig. 1. The model provides a robust framework for modelling intelligent agents by simulating human-like reasoning and decision-making. Particular focus is placed on key aspects such as goal-based plan generation, intention management, and dynamic environment integration (beliefs). One potential advantage of the BDI agent architecture for this approach is its ability to recover from plan failures and attempt alternative strategies, reducing the impact of predictable LLM failures on the agent's behaviour.

For this paper, we are specifically targetting the generation of plans for an already-active BDI agent. In Jadex, BDI agents are normally provided with a set of plans implemented for example as Java classes or methods. Our goal is to allow an agent to execute without a predefined plan, using the an LLM to generate the plan for the agent at runtime.

To achieve this, it is necessary to address a number of critical issues in a comprehensive manner. First, the approach needs to be implemented and integrated into the Jadex BDI architecture in a way that is non-invasive and ensures compatibility with the Jadex BDI execution engine. Since the goal is to minimise the programming effort required for plan implementation, the agent should ideally be able to achieve its goals without being provided a plan at all, executing simply based on its goals, context, beliefs and static information available. This information needs to be prepared for the LLM in a manner that the chosen LLM is capable of processing and then use to provide executable code for a plan. Choosing an appropriate LLM also requires consideration of factors such as model size, capabilities, and possible

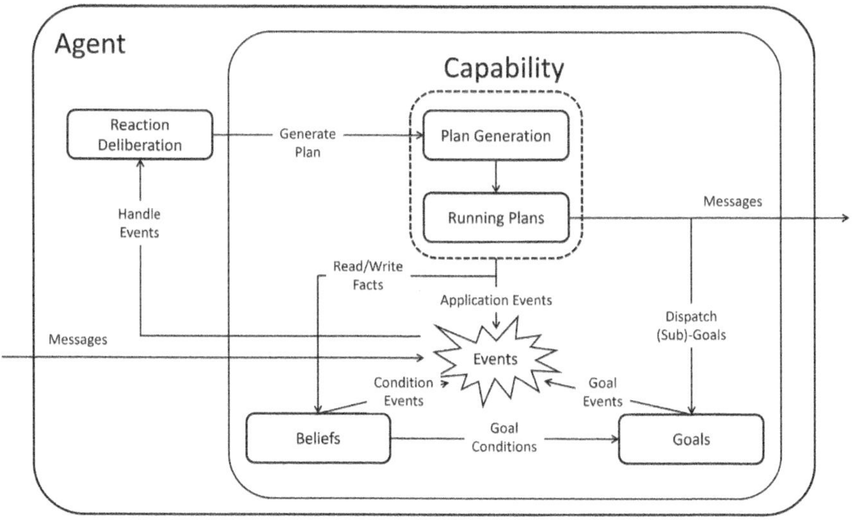

Fig. 1 Jadex abstract architecture with transformer-based plan generation approach

specialization, for example in code generation. While it may initially seem obvious to choose a specialized LLM for code generation, it is not necessarily the case since the LLM also needs to process a lot of non-code information and successfully use it to generate the code. The selection process involves identifying a transformer model that is capable of effectively generating plans while maintaining uniformity in both the prompt and extension mechanisms. An important aspect of this integration process is the extension of the Jadex Architecture with a novel method, which will facilitate seamless integration and enhance the overall functionality of the system.

The integration of LLM into this architecture enhances the agent's ability to handle complex and dynamic environments through advanced natural language understanding and generation. It is of note that while implementation aspects and execution details have changed over time, for example by enabling pure Java-based BDI agents instead of XML, the fundamental structure and approach of the BDI architecture in Jadex, as presented in [4], remains, in general, the same. The goal of this approach is to target the section of plan execution, where instead of executing a predefined and implemented plan, the goal is to implement an approach that incorporates the capabilities of an LLM to generate the plan on the fly.

The generation of plans in Jadex is typically achieved through a plan library approach, whereby plans are implemented as Java classes and stored in a library. This approach facilitates the efficient reuse and modularisation of plans, thereby promoting a structured development of agent behaviour. However, by integrating a LLM, the planning mechanism can be enhanced to dynamically generate plans without the need for predefined programming, allowing agents to autonomously formulate actions based on real-time data and contextual analysis. This advancement significantly increases the flexibility and adaptability of the agent system, enabling more sophisticated and responsive behaviour. Intention management in Jadex is closely tied to the agent's goals. The framework enables agents to dynamically formulate, pursue and adjust goals based on changing environmental conditions. This flexibility allows them to adapt their behavior in response to new information and objectives [17].

As illustrated in Fig. 1, the Jadex architecture regards agents as black-box systems that respond to external events, such as messages or goal events. Beliefs are represented in an object-oriented manner and are actively involved in the agent's execution, reacting to changes in the belief state. Goals are a central aspect of Jadex, featuring a lifecycle that includes options, active, and suspended states.

Plans, representing the behavioural aspect of the agent, are selected and executed based on events, goals, and contextual conditions. Capabilities group elements such as beliefs, goals, and plans into reusable modules that encapsulate specific functions and can be connected with other modules via flexible import/export mechanisms [4].

This study presents a new method for generating plans in agent systems, using a Transformer model connected via HTTP request. As an initial step in the development process, we focused on integrating LLM-based plan generation into Jadex through the agent application. This integration served as a crucial preliminary assessment of the Transformer model's suitability for our requirements and the agent's domain. Designing the agent as planless ensures that the basic building blocks for creating the agent remain unchanged. However, minor adjustments are necessary for the integration of the Transformer model. Building blocks in form of support classes have been developed to facilitate communication between the agent and the Transformer model, process the plans generated by the model and dynamically compile these plans into executable code that can be executed by the agent (see Fig. 2).

4.1 Providing Context Information

In order for the transformer model to be able to generate useful and executable code, it needs to be provided with context information. The most important context information is the API of the code the plan needs to invoke. For example, if a plan needs to update information in a customer object, the transformer model needs to be made aware of the class and methods of the customer object, so it can generate code that invokes, for example, the right method to change the information.

As a result, information about the class and API structure of ancillary classes that are important to the plan needs to be extracted and provided to the transformer model in a usable fashion.

In order to extract the class and API information, we process the relevant classes before the plan generation prompt. A simple approach would be to use a reflection-based approach, using the Java Reflection API to extract information about the classes and methods involved in the application context. However, since this requires Java to load and process all of such classes, it is a relatively slow processes. Since the system is simply targets all classes potentially involved in the plan, it may require Java to

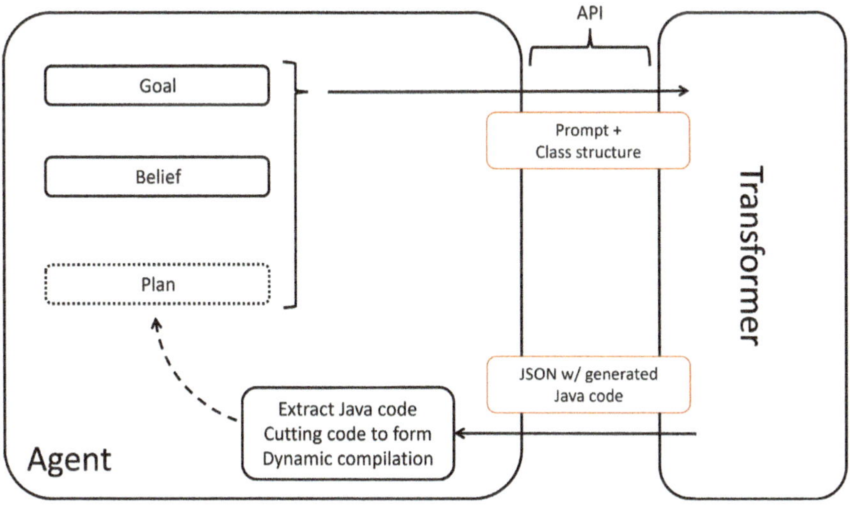

Fig. 2 Approach used to generate plans for BDI agents using a LLM

load classes that ultimately turn out to be irrelevant to the plan by the transformer model.

Therefore, the chosen approach extracts only the minimal information required directly from the Java bytecode contained in the class files, parsing the binary content speed up the extraction process.

In comparison to static class extraction or manual inspection, our real-time approach offers several advantages: the agent automatically extracts the required information when generating a plan and adapts to changes in the application classes. This ensures that the transformer model always receives up-to-date information and avoids generating plans for APIs that no longer exist or have changed. Moreover, in contrast to manual inspection, which is both time-consuming and prone to error, our automated approach ensures efficiency and accuracy in class information extraction.

The class information is then processed and transmitted to the LLM by compiling the information into a prompt, which facilitates the generation of plan classes that are seamlessly integrated with the agent system's structures. The model uses the provided class information to ensure coherence and adaptability of the generated plans to the agent application classes.

4.2 LLM Prompt Engineering

While the code generation capabilities of LLMs are generally impressive, they do have their limitations in certain aspects. In particular, LLMs start to struggle if provided with a very large context with a lot of API details that need to be considered when generating the code. The more the LLM needs to consider the correct use of annotations, methods, classes and parameters, the more likely it is to hallucinate on some detail and generate wrong or even non-compilable code.

In order to reduce the contextual load for the LLM when generating the plan code, the LLM is not asked to generate a Jadex BDI plan class directly, since this would not only require the LLM to consider the application classes of the BDI agent but also the whole of the Jadex BDI agent API, increasing the chance of the LLM making a mistake. Furthermore, the predefined plan class currently also contains the generic code for prompt generation, invocation of the LLM, compilation of the generated code and invocation (see Fig. 3).

As a result, the implemented LLM prompt engineering process only requires the LLM to generate the "active" code of a plan by generating the prompt to generate a generic, non-BDI, non-Jadex 'Planstep' class with a single 'doplanstep' method, which receives its context object using its parameters. The system will then generate a generic, predefined plan class that instantiates and invokes the Planstep class as part of its plan method.

This indirect approach shields the LLM from knowledge about the Jadex API, avoiding some potential mistakes the LLM could make. This approach was refined through an experimental investigation employing a trial-and-error approach. The initial step involves the formulation of a predefined prompt, which specifies the

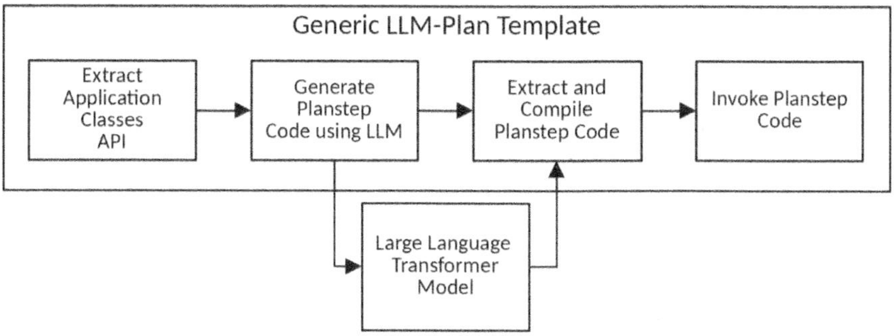

Fig. 3 Generic BDI plan invoking the LLM to generate code

generation of code blocks to implement logic within the 'Planstep' class. This prompt serves as a guideline for the LLM to generate code snippets that fulfil the specified requirements.

As mentioned, the prompt includes all necessary context information about the application classes and API as well as a goal description, relevant beliefs and asks the LLM to generate an appropriate piece of code to solve the problem based on the goal description. We specifically limit the LLM's knowledge that it is part of a BDI agent system and the Jadex API specifically to narrow its focus on solving the problem at hand without getting distracted by other aspects of the system.

The provided class structures, including the 'infostring' consisting of class information and the input prompt, serve as guidelines for the generated Java code. This code is extracted and further processed for evaluation. The generated output can then be processed and used for integration into the agent.

4.3 Code Compilation and Execution

The extraction phase commences upon the LLM response, which is typically formatted in JSON. This structured format presents code snippets alongside metadata, which facilitates comprehension and processing. Parsing the JSON alone, however, is insufficient. The response of the LLM itself needs to be further processed as well, for example by identifying and isolating relevant code blocks in accordance with the prompts. This is because the output of the LLM can vary and the code may be embedded in further explanatory text generated by the LLM. This text may appear even if the LLM was specifically instructed to omit it.

Once the code of the doplanstep method has been extracted, it is embedded in a template Planstep class in-memory as a Java String. The built-in Java compiler from the "java.compiler" module of the Java API is then used to compile the generated class source code to Java byte code. This byte code is then injected into a ClassLoader instance to generate a usable Java class completely in-memory and generate an instance of the Planstep object.

In the context of user interaction within an agent-based system, the process of code compilation and execution plays a fundamental role in the realisation of the intended functionality of generated plans. Upon reaching the plan annotation that has been triggered by a goal annotation, the agent initiates the plan execution process. A central aspect of this process is the verification of the existence of the generated plan, which serves to ensure readiness for subsequent phases.

A series of steps is employed to facilitate the compilation and execution of the generated code. Initially, the provided class structures, including the 'infostring' consisting of class information and the input prompt, serve as a framework for the generated Java code. Subsequently, the code is translated into an executable Java class utilising standard Java libraries. It is noteworthy that the runtime compilation process generates a 'Planstep.class' file, which encapsulates the logic specified in the generated code, on a dynamic basis.

5 Example Application and Evaluation

The utilisation of a decoder transformer in the example application demonstrates significant utility due to its inherent capability to generate sequential outputs based on input data. Unlike traditional planning approaches that rely on predefined libraries, the use of a decoder transformer facilitates dynamic and adaptable planning processes. The transformer's self-attention mechanism enables the decoder to effectively capture complex dependencies within the input data, thus facilitating more nuanced and context-aware decision-making. Furthermore, the decoder's capacity to generate output sequences in an iterative manner aligns well with the trial-and-error methodology, allowing the agent to refine its plans iteratively through interaction with its environment. This approach not only enhances the agent's flexibility and adaptability but also adheres to scientific rigour, thereby enabling empirical evaluation and refinement of the planning process.

In this experimental setup, an agent was developed without statically defined plans. Instead, the agent is structured based on its Beliefs and Desires, with its Intentions being generated dynamically by an AI model. The experiments were intentionally kept simple to conduct an initial assessment of whether this approach holds potential. The goal was to determine the feasibility of the agent's ability to operate effectively and adaptively in simple scenarios, with the future outlook to apply this approach in more complex environments and explore different solution strategies. These initial experiments also aimed to identify potential challenges and limitations in the system.

To support this evaluation, a comparative analysis was conducted between GPT-3.5, Vicuna-7b, and CodeLlama-7b to assess whether a smaller model with self-hosting capabilities could meet the intended requirements. The models were tested for their coding proficiency, with a focus on error rates and system reliability. Using the most recently generated prompts, we were able to evaluate the robustness and dependability of the developed agent across a range of diverse scenarios.

The results indicate that the GPT model produced the expected outcome in approximately 84% of the analyzed cases. However, in the remaining cases, deviations from the expected behaviour were observed. The self-hosted models, such as the Vicuna system, achieved the target in 8% of cases, while the CodeLlama model was successful in 40% of cases. This highlights a significant performance difference between the two systems.

Analyzing the results further (refer to Fig. 4), patterns emerge from the analysis of the respective behaviours of Vicuna and CodeLlama when comparing their performance. Vicuna successfully met the required criteria on eight occasions but encountered challenges in 22 instances, resulting in the return of unsorted lists due to implementation shortcomings. Additionally, there were 65 compilation errors, mainly caused by code generation errors related to data types and enum values. In contrast, CodeLlama complied with the specifications in 40 cases, with only two instances of returning unsorted lists. However, similar to Vicuna, compilation errors persisted in 53 instances, indicating a common issue across both systems. These findings emphasize the significance of meticulous implementation and robust code generation processes to ensure the effectiveness and dependability of automated systems like Vicuna and CodeLlama.

In our application of GPT for code generation, we have encountered some challenges. Specifically, we have found that around 3% of the generated code contains inaccuracies that result in compilation errors. Additionally, approximately 13% of the produced lists remain unsorted due to implementation deficiencies.

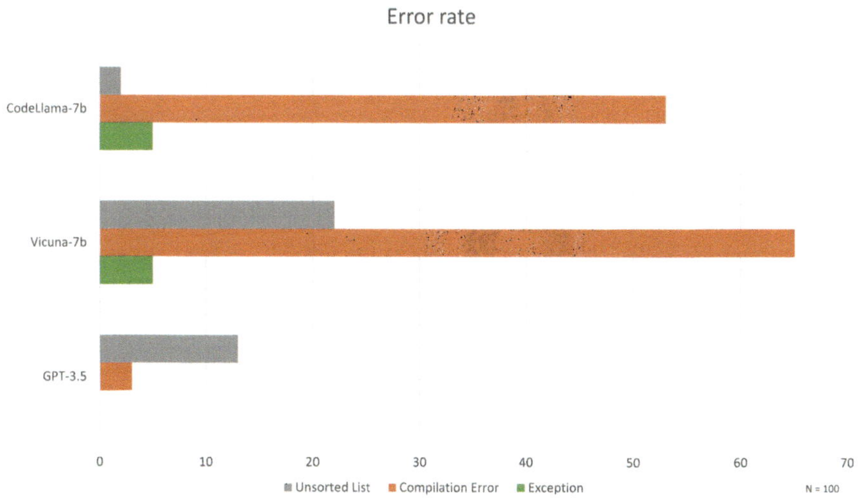

Fig. 4 Model error rate for GPT 3.5, Vicuna-7b and CodeLlama-7b

The test results reveal a notable difference in the behaviour of GPT and the self-hosting models regarding logic implementation. GPT generates comments more frequently, suggesting the need for logic implementation at a specific point in the code. In contrast, Vicuna and CodeLlama appears to implement some form of logic in most cases, regardless of its correctness or relevance to the task. This observation highlights that Vicuna may implement logic without the necessary precision and focus on the specific input requirements. Additionally, GPT has greater error resistance, particularly when dealing with unimportant or irrelevant class parameters. This suggests that GPT is better at identifying and focusing on the essential elements of an input while disregarding unimportant or misleading information. GPT's potential for robustness and versatility is increased by this property, particularly in situations where input data is imperfect or requires interpretation.

In comparison to the current state of the art, this approach is both functional and more intuitive than existing planning methods, as it does not require extensive programming or the provision of specific planning algorithms and plan libraries. By leveraging Transformer models and the BDI architecture in Jadex, agents can dynamically generate and adapt plans at runtime without the need for manually coded plans. This significantly reduces the complexity and effort involved in developing autonomous agents, making the process more accessible and efficient.

6 Conclusions and Future Work

The current understanding of the functionality of autonomous AI agents, such as Agent-GPT and Baby AGI, is limited. However, Jadex, with its defined BDI architecture that integrates Goals, Beliefs and Plans, provides a robust foundation for the development and extension of these technologies. The combination of Java, a stable programming language with defined rules and types, and Transformer models could represent a significant step towards the explainability of these systems.

Although the presented work is an initial attempt to demonstrate feasibility of integrating LLM and agent technology, it opens up avenues for the development of more complex systems.

First, the current approach could be further refined. For example, the BDI agent and its plan are only as smart as the underlying transformer model. Since these models, as statistical systems, are not yet able to operate completely autonomously, future research should focus on exploring a larger set of LLMs and the potential of combining LLMs for the automatic generation of plans. There are several avenues for future improvement of the LLM prompt engineering process. These include the exploration of alternative prompt formulations, the integration of additional contextual information, and the optimisation of the iterative refinement process. In addition, the potential to further improve the effectiveness and efficiency of LLM prompt engineering can be realised through the application of advanced machine learning techniques, such as reinforcement learning and transfer learning. This approach focuses on sustainability by prioritising smaller models that can conserve resources and be hosted locally, enabling secure processing of potentially sensitive data. When integrated with the BDI framework, these smaller models can produce results comparable to larger models. As a futher refinement, successful plans could also be stored in an automated plan repository to be used directly in the right context or further adapted by LLMs.

Second, other parts of the BDI architecture could be targeted using LLMs or other machine learning approaches: Multiple plans could be generated or provided by a human developer in a hybrid system, with an LLM or other machine learning system assisting on the means-end reasoning level, making choice and priotizing plans given the currently active set of goals. On the goal level, two more options open up: The goal deliberation stage, which decides which goals are actively pursued over others could be assisted by a machine learning system. Finally, goals themselves could be dynamically decomposed into subgoals using a pure machine learning or human-machine hybrid system, ultimately only requiring the initial agent design to provide context, environment and a limited set of general goals to pursue, that will be automatically or semi-automatically decomposed and then executed using generated plans.

In this regard, this paper has provided an initial promising foray into the use of the BDI architecture as a structural framework for a sort of modular AI system, where the machine learning components will provide the flexibility and the BDI agent provides the structure and explainability (using its goals) of the overall system.

References

1. Michael Bratman. *Intention, Plans, and Practical Reason*. Cambridge, MA: Harvard University Press, Cambridge, 1987.
2. Rafael H Bordini, Amal El Fallah Seghrouchni, Koen Hindriks, Brian Logan, and Alessandro Ricci. Agent programming in the cognitive era. *Autonomous Agents and Multi-Agent Systems*, 34(2):37, May 2020.
3. Rafael H Bordini, Jomi Fred Hübner, Michael Wooldridge. Programming multi-agent systems in AgentSpeak using Jason. John Wiley & Sons, 2007
4. Lars Braubach, Alexander Pokahr, and Winfried Lamersdorf. Jadex: A short overview. In *Main Conference Net. ObjectDays*, volume 2004, pages 195–207. AgentExpo, 2004.
5. Rafael C Cardoso and Angelo Ferrando. A review of agent-based programming for multi-agent systems. *Computers*, 10(2):16, 2021.
6. Rafael C Cardoso, Angelo Ferrando, and Fabio Papacchini. Automated planning and bdi agents: A case study. In *International Conference on Practical Applications of Agents and Multi-Agent Systems*, pages 52–63. Springer, 2021.
7. Lavindra De Silva, Felipe Rech Meneguzzi, and Brian Logan. Bdi agent architectures: A survey. In *Proceedings of the 29th International Joint Conference on Artificial Intelligence (IJCAI), 2020, Japão.*, 2020.
8. Lavindra De Silva, Sebastian Sardina, Lin Padgham. First principles planning in BDI systems. AAMAS (2), 1105–1112, 2009
9. Jessica L Espejel, Mahaman Sanoussi Yahaya Alassan, El Mehdi Chouham, Walid Dahhane, and El Hassane Ettifouri. A comprehensive review of state-of-the-art methods for java code generation from natural language text. *Natural Language Processing Journal*, 3:100013, 2023.
10. Significant Gravitas. Autogpt, 2023. GitHub repository v0.5.0.

11. Andreas Herzig, Emiliano Lorini, Laurent Perrussel, and Zhanhao Xiao. Bdi logics for bdi architectures: old problems, new perspectives. *KI-Künstliche Intelligenz*, 31:73–83, 2017.
12. Koen V Hindriks and Tijmen Roberti. Goal as a planning formalism. In *German Conference on Multiagent System Technologies*, pages 29–40. Springer, 2009.
13. Guohao Li, Hasan Abed Al Kader Hammoud, Hani Itani, Dmitrii Khizbullin, and Bernard Ghanem. Camel: Communicative agents for "mind" exploration of large language model society, 2023.
14. Zelong Li, Wenyue Hua, Hao Wang, He Zhu, and Yongfeng Zhang. Formal-llm: Integrating formal language and natural language for controllable llm-based agents. arXiv preprint arXiv:2402.00798, 2024.
15. Felipe Meneguzzi and Lavindra De Silva. Planning in bdi agents: a survey of the integration of planning algorithms and agent reasoning. *The Knowledge Engineering Review*, 30(1):1–44, 2015.
16. Joon Sung Park, Joseph O'Brien, Carrie Jun Cai, Meredith Ringel Morris, Percy Liang, and Michael S Bernstein. Generative agents: Interactive simulacra of human behavior. In *Proceedings of the 36th annual acm symposium on user interface software and technology*, pages 1–22, 2023.
17. Alexander Pokahr, Lars Braubach, and Winfried Lamersdorf. Jadex: A bdi reasoning engine. *Multi-agent programming: Languages, platforms and applications*, pages 149–174, 2005.
18. Anand S Rao, Michael P Georgeff, et al. Bdi agents: from theory to practice. In *Icmas*, volume 95, pages 312–319, 1995.
19. Reworkd. Agentgpt, 2023. GitHub repository v.1.0.0.
20. Sebastian Sardina, Lavindra De Silva, and Lin Padgham. Hierarchical planning in bdi agent programming languages: A formal approach. In *Proceedings of the fifth international joint conference on Autonomous agents and multiagent systems*, pages 1001–1008, 2006.
21. Andrea Traldi, Francesco Bruschetti, Marco Robol, Marco Roveri, and Paolo Giorgini. Real-time bdi agents: a model and its implementation. arXiv preprint arXiv:2205.00979, pages 511–517, 2022.
22. A Vaswani. Attention is all you need. *Advances in Neural Information Processing Systems*, 2017.
23. Andrzej Walczak, Lars Braubach, Alexander Pokahr, and Winfried Lamersdorf. Augmenting bdi agents with deliberative planning techniques. In *Programming Multi-Agent Systems: 4th International Workshop, ProMAS 2006, Hakodate, Japan, May 9, 2006, Revised and Invited Papers 4*, pages 113–127. Springer, 2007.
24. Lei Wang, Chen Ma, Xueyang Feng, Zeyu Zhang, Hao Yang, Jingsen Zhang, Zhiyuan Chen, Jiakai Tang, Xu Chen, Yankai Lin, et al. A survey on large language model based autonomous agents. *Frontiers of Computer Science*, 18(6):186345, 2024.
25. Mengwei Xu, Kim Bauters, Kevin McAreavey, and Weiru Liu. A framework for plan library evolution in bdi agent systems. In *2018 IEEE 30th International Conference on Tools with Artificial Intelligence (ICTAI)*, pages 414–421. IEEE, 2018.
26. Hui Yang, Sifu Yue, and Yunzhong He. Auto-gpt for online decision making: Benchmarks and additional opinions. arXiv preprint arXiv:2306.02224, 2023.
27. Li Zhong and Zilong Wang. A study on robustness and reliability of large language model code generation. arXiv preprint arXiv:2308.10335, 2023.

Leveraging LLMs for Smart Cities Qualitative Data Analysis

Elisa Covato, Kamran Soomro, Zaheer Khan, Muhammad Bilal, Sam Kirby, and Tom Yiangou

Abstract Public authorities frequently conduct surveys and analyse data from citizens, a process that is often labour-intensive when performed manually. This paper explores how Generative Artificial Intelligence (GAI) can assist in automating data analysis for public authorities. In this respect, we investigate the potential of Large Language Models (LLMs) to perform sentiment analysis and summarisation of unstructured data as smart services. Using data from the East Bristol Liveable Neighbourhood (EBLN) as a case study, we assess the accuracy and precision of these models and validate the results against ground truth data and expert evaluations. Our findings indicate that sentiment classification achieved over 90% accuracy. In contrast, the expert validation rated the summarisation without context accuracy highly satisfactory, however, the satisfaction was low when contextual summarisation was evaluated. These results suggest that LLMs offer a promising approach to improving the efficiency of qualitative analysis, though further research is required to enhance their accuracy and usefulness.

Keywords Sentiment analysis · Text summarisation · Smart cities · Generative Artificial intelligence

1 Introduction

Cities are undergoing significant transformations as they seek to harness the full potential of urban governance driven by Information and Communication Technology [1]. Several benefits are sought to manage the complexity of urban systems

E. Covato (✉) · K. Soomro · Z. Khan
School of Computing and Creative Technologies, University of the West of England, Bristol, UK
e-mail: elisa.covato@uwe.ac.uk

M. Bilal
Business School, Birmingham City University, Birmingham, UK

S. Kirby · T. Yiangou
Bristol City Council, Bristol, UK

N. Polatidis et al. (eds.), *Intelligent Distributed Computing XVII*, Studies in Computational Intelligence 1203, https://doi.org/10.1007/978-3-031-87639-4_16

and gain insights from public participation in planning processes [6, 18]. An important driving force in this transformation concerns the information overload from urban systems and local initiatives such as citizen observatories. The recent policy shift towards more emphasis on citizen engagement in data gathering, planning and decision-making processes e.g., European Bauhaus,[1] Built4People[2] and EU Mission Cities,[3] requires further technological support to manage and analyse such data. In particular, much of the data is gathered through surveys or comments provided by citizens on specific plans or topics via social media or other participation platforms. Since these data are typically unstructured, analysing it can be time-consuming and requires specific skills. This problem can be exacerbated by the growing number of surveys and high levels of public participation.

Over the past decade, data analytics has emerged as an important innovative technology in the domain of smart cities [1, 22]. Various statistical, machine learning and data mining techniques have proven to be revolutionary in the context of Smart City Data Analytics. Among these, deep learning models have become one of the most cutting-edge technologies, offering significant contributions across a range of applications for smart city services [15, 26].

Recently, disruptive technologies such as Generative Artificial Intelligence (GAI) and Large Language Models (LLMs) are revolutionising the way data analysis can be performed. Their benefits cut across domains such as transportation, urban planning, and smart buildings. Their ability to analyse and evaluate unstructured qualitative data with high accuracy and precision makes them ideal candidates to help public authorities streamline their processes. In this respect, we investigate the suitability of LLMs for two automated services for public authorities: (i) sentiment analysis and (ii) summarisation for survey data. We applied selected models to real, unstructured textual data collected through the East Bristol Liveable Neighbourhood (EBLN)[4] initiative in Bristol, United Kingdom. The EBLN is one of the pilots under the Horizon Europe GREENGAGE project. [5]The GREENGAGE project aims to develop citizen observatories to facilitate European GREEN Deal policy objectives across different pilots in EU and UK. By leveraging citizen science, the project will collect extensive unstructured qualitative data through surveys, discussions and comments gathered through the GREENGAGE GREEN Engine, such as mobile apps. Additionally, the project will combine sensor data from traffic monitoring and pollution levels, with citizens' feedback to raise public awareness and engagement in environmental issues, as well as to gather their opinions and needs on these topics.

The remainder of this paper is structured as follows: Sect. 2 presents background information on LLMs and their suitability for data analysis tasks. Section 3 introduces the EBLN dataset, detailing our pre-processing steps, model architectures, and

[1] https://new-european-bauhaus.europa.eu/about/about-initiative/_en.

[2] https://built4people.eu/.

[3] https://netzerocities.eu/mission-cities/.

[4] https://eastbristolliveableneighbourhoods.commonplace.is/.

[5] https://www.greengage-project.eu/.

experimental design of the two key services: sentiment analysis and text summarisation. We discuss our findings and their implications in Sect. 4, followed by conclusive remarks in Sect. 5.

2 Background

2.1 Large Language Models for Qualitative Analytics

Traditional approaches to qualitative data analysis rely on understanding vast amounts of unstructured data (text, video, or audio) through manual processes [8]. These approaches typically require significant time investments, and are heavily dependent on human expertise, inevitably incorporating individual biases into the analytical process.

The rise of Large Language Models has encouraged a shift towards developing more sophisticated, automated systems for qualitative data analysis. These LLMs, comprising billions of parameters and trained on vast datasets, have demonstrated remarkable capabilities in various language processing tasks [16]. Various studies have explored the application of LLMs in qualitative analysis, focusing on specific tasks such as sentiment analysis [13], text summarisation [2, 27], theory-driven categorisation [25], or holistic solutions [19, 23]. However, it remains a gap in understanding their practical benefits in real-world scenarios. Current research often lacks evaluation from end-users, or applications in real smart cities case studies. This study aims to address this gap by applying LLMs on recent survey data, and incorporating feedback from actual practitioners. Our final goal is to set the foundation for future research directions, and assess the true potential of LLMs in qualitative data analysis within practical contexts.

2.2 Zero-Shot Learning and LLMs

Training machine learning models is a resource-intensive task, requiring substantial data, time, and computational power [20]. However, real-world scenarios often impose limitations on these resources. Zero-shot learning is a machine learning paradigm that enables models to make predictions for classes they have not encountered during training [3], reducing computational costs and the need for large amounts of annotated datasets. While this has been in the past mainly used for computer vision, it has recently gained attraction with the surge of LLMs [24].

LLMs are pre-trained on vast amounts of data, capable of performing a wide range of language tasks effectively. Zero-shot learning leverages this extensive knowledge base, making LLMs flexible and adaptable across various scenarios. Given the ability of LLMs to understand and generate human-like text, LLMs can be directed to perform zero-shot learning simply through natural language instructions or 'prompts' [14]. This conversational manner of task description allows these models to adapt to new contexts without the need for task-specific training data, time-consuming data preparation steps or extensive model fine-tuning. Moreover, this approach is particularly beneficial in providing non-experts with a tool to perform language analysis.

In this study, we have created suitable prompts to perform zero-shot learning for sentiment classification and summarisation. Creating effective prompts involves a trial-and-error approach, which can be challenging [17]. While exploring the nuances of prompt creation is outside the scope of this initial study, we will focus on discussing the results obtained. It is important to note that varying prompts may yield different results in this context.

2.3 Sentiment Classification Using LLMs

Understanding public opinion, citizens' experiences and satisfaction is crucial in the context of Smart Cities. This enables authorities to assess initiative reception, pinpoint concerns, improve quality of life, and ensure responsive governance. To accomplish these objectives effectively, sentiment classification emerges as an invaluable analytical tool [7]. This natural language processing technique aims to determine the emotional tone or attitude expressed in a piece of text. It typically involves categorising text into binary sentiments—positive or negative—although more nuanced classifications are also feasible.

Traditional sentiment classification techniques often rely on lexicon-based methods, which use predefined dictionaries of sentiment-bearing words, or machine learning approaches such as Naive Bayes, Support Vector Machines and Recurrent Neural Networks (RNNs) which learn from labelled data [21]. These techniques, while effective, can struggle with context-dependent sentiments, require large amounts of labelled data for training, demand extensive feature engineering and dataset preparation, which can be time-consuming and costly to perform. In contrast, LLMs excel in understanding context and nuanced language, require minimal task-specific training due to their pre-trained nature, and can perform well with limited or no labelled data [13, 21]. LLMs' versatility in processing sentiments across a range of domains, with superior accuracy, makes them ideal for sentiment analysis, especially in the context of survey data collection, characterised by large quantitative of unstructured data covering different topics.

2.4 Flan-T5: Model Selected for Sentiment Classification

For the sentiment classification task in this study, we have tested Flan-T5 language model, developed by Google AI in 2022 [4]. Flan-T5 is an open-source, improved version of the original T5 (Text-to-Text Transfer Transformer) model, treating all natural language processing tasks as text-to-text problems. This model has been fine-tuned on a diverse range of tasks using instruction-based learning, making it suitable to be used in zero-shot learning context. This versatile approach adapts well to sentiment classification, with the model's instruction following ability making it an easy tool for language analysis, even for non-technical users. Flan-T5 is comparable in performance to many heavyweight, proprietary models while being open-source [28], reducing cost constraints and allowing customisation. These features make it an excellent candidate for this investigation, balancing advanced capabilities, flexibility, and cost-effectiveness. Google has released multiple variants of Flan-T5, ranging from compact models with 80 million parameters and 300 MB in size, to expansive versions with 11 billion parameters requiring 80 GB of memory. For this study, we evaluated the following checkpoints available on Hugging Face[6]

1. Flan-T5-small: 80 million parameters, 300 MB
2. Flan-T5-base: 250 million parameters, 990 MB
3. Flan-T5-large: 780 million parameters, 1 GB.

The choice of these model sizes was motivated by the aim to assess the performance on common laptops with limited resources, evaluate the trade-off between execution time and accuracy across different model sizes, and make the approach accessible to a wide range of users.

2.5 Summarisation Using LLMs

In the digital age, where information is abundant and readily available, the ability to condense and summarise text is of high importance. Text summarisation serves as a powerful tool that aids in the efficient extraction and comprehension of key information from large volumes of data. It enables individuals and businesses to sift through vast amounts of content and quickly identify the most relevant and critical information. By transforming lengthy documents into concise summaries, it saves time, enhances productivity, and supports informed decision-making processes. As such, text summarisation plays a crucial role in navigating information overload in today's data-driven world.

Text summarisation approaches have been around for a while. For example, frequency-based, graph-based and topic-based approaches aid in extractive summarisation techniques. On the other hand, Sequence-to-Sequence models such as RNNs and LSTMs aid abstractive summarisation. The emergence of LLMs promises to usher in a new era of summarisation techniques. However, being relatively recent, not much work has been performed to investigate the performance of LLMs for summarisation in various domains. Although there is limited research, it suggests that traditional approaches still perform reasonably well for generic tasks, though LLMs

[6] https://huggingface.co/docs/transformers/main/en/model/char‘_doc/flan-t5.

might outperform them when fine-tuned for specific domains [9, 10]. Therefore, in this paper, we explore how LLMs perform in summarisation of survey data.

2.6 Llama3: Model Selected for Summarisation

For the summarisation service, we have chosen the Llama3 models. This model has been recently released by Meta in 8B and 70B parameter scales in both instruction-tuned and fine-tuned varieties.[7] We opted to test Llama3 because it is a new model that has shown promising results in various tasks, including summarisation, with comparable or superior performance to other leading LLMs.

The LLama3 models have been trained using a variety of techniques to optimise both the pretraining and post-training steps. In particular, a dataset containing 15 trillion tokens was specifically curated from publicly available sources to train the model. Rigorous quality control measures, including heuristic filters and eliminating semantic duplicates, have been implemented to improve data reliability compared to previous Llama versions. Taking into account these factors, along with the model's evaluation of 1,800 prompts across 12 essential tasks, including summarisation, we have selected this model specifically for our data summarisation task with the aim to achieve robust, efficient, and practical summarisation results.

3 Methodology and Implementation

3.1 Case Study and Data Collection

The East Bristol Liveable Neighbourhood (EBLN) project is a community-driven initiative to transform East Bristol's streets. The main goal is to collaborate with local residents, workers, students and commuters to design streets that are liveable and people-friendly. This project is part of a co-design phase aimed at developing long-term solutions with the help of the local community. The EBLN is also a pilot in the Horizon Europe GREENGAGE project, which promotes inclusive citizen observatories by using digital solutions.

The survey data used in this study were collected between January and March 2022 by Bristol City Council, UK. Participants, including those living, working, or travelling to the area, shared their opinions either using an interactive online map,[8] or by engaging with members of Bristol City Council through door-to-door interviews or community discussions.

[7] https://ai.meta.com/blog/meta-llama-3/.

[8] https://eastbristolliveableneighbourhoods.commonplace.is/map/map.

Face-to-face interviews were conducted and recorded manually. Citizens were asked about their thoughts on what they appreciate about their local area, and how they would like to improve it. They had the option to respond to either or both questions, resulting in over 400 individual contributions and a total of 845 comments of varying length. While the complete dataset cannot be disclosed due to confidentiality constraints, Table 1 presents a representative sample of responses collected.

Online, citizens could express their feedback by:

1. Marking specific locations on a map
2. Linking a feeling, chosen from a range of five sentiments, spanning from negative to positive, with the selected area
3. Providing optional comments in a free-text box
4. Additionally, selecting one or more topics related to their comments, reasons for their expressed feeling, and suggestions for improving the area

An example of response from the online survey is shown in Table 2. The dataset collected through the online tool comprises 540 geo-located entries, of which 90% contains both textual comments and sentiment labels. A preliminary study conducted on this dataset [5], which involved analysing the textual comments alongside the provided sentiment and topic labels, showed that most participants lived within the area and tended to emphasise negative aspects of their neighbourhood, particularly in relation to main roads and junctions. Interestingly, the word "road" frequently appeared in negative contexts, while "street" was associated with positive sentiments, reflecting nuanced perceptions of different urban spaces. Conversely, areas with green spaces generally received more positive feedback. Sentiment analysis showed that negative statements were more prevalent around safety and environment-related comments, while texts linked to community aspects displayed a more balanced distribution of sentiments.

The preliminary study [5] demonstrated the potential of analysing public feedback to inform urban planning initiatives. However, it also highlights the need for scalable and efficient methodologies. While the EBLN dataset has labelled textual data, most urban datasets lack such structured annotations, resulting in the need for extensive time to analyse and generate comprehensive summaries to extract insights and actionable steps.

Table 1 Example of responses from the EBLN interviews

Q1: What do you like about your local area?
A1: Good sense of community A2: It's fairly diverse—there's a real sense of community, it feels very friendly A3: Park, shops
Q2: How would you like to make it better?
A4: More activities for children and improve parks. Homes are too old and damaged and we need repairs A5: Traffic A6: Wider roads

Table 2 Example of responses from the online EBLN map

Sentiment	Positive	Negative
Subjects	Trees and greenery on street, Street trees and planting	Walking, Crossings
Reasons	Pleasant	Not pedestrian friendly, Difficult to cross the street
Suggestions	Slow down traffic	Add crossing, Safer junction for walking and cycling
Comments	The street planters have reduced traffic speed and made the street'greener'. Something similar could be done in other locations within the project area and in traffic displacement areas out-side the project area	Hard to cross here - there is a traffic island slightly above this point but often want to cross lower down and it's hard to do so as the road is busy with two lanes of fast traffic

Table 3 Sentiment classification prompt

```
Please perform Sentiment Classification task.

Given the sentence, assign a sentiment label from
['negative','positive'].
Return label only.

Sentence: {input_text} Label:
```

Table 4 Summarisation system prompt 1

```
Provide a detailed summary of the following document
using the given context. Be as helpful as possible.

context: "{context}"
document: "{text}"
```

Table 5 Summarisation system prompt

```
Summarise the provided document, using the supplied
context if provided.

context: "{context}"
document: "{text}"
```

In this study, we have tested our LLM-based methods against the EBLN datasets. For sentiment classification, we have focused on the online EBLN dataset, leveraging its pre-existing sentiment labels to assess our approach's accuracy and effectiveness. For the summarisation tasks, we have worked with both the online and in-person collected data, and results have been evaluated by members of Bristol City Council directly involved in the EBLN and GREENGAGE project.

3.2 Data Preprocessing

Our work centers exclusively on the free-text comments, and their corresponding sentiment labels in the case of the sentiment classification task, therefore all other features in our datasets were excluded. In line with our aim of utilising LLMs, our preprocessing was intentionally minimal. This is because LLMs are designed to handle natural language inputs, including variations in spelling, grammar, and informal expressions commonly found in public feedback. Therefore, for both tasks, we retained all comments regardless of length to preserve the full range of citizen feedback, since even brief comments can provide insight into urban perceptions.

For the summarisation tasks, two files were generated that included all the text comments. These comments were either placed in one column for the online dataset or in two columns for the two questions asked during the interviews. In the case of the sentiment classification task, we performed a sentiment simplification. The original dataset included five sentiment categories: positive, mostly positive, neutral, mostly negative, negative. Following the approach in [5] we have excluded the neutral comments, and grouped the remaining labels into two polar groups, positive and negative. We simplified the sentiments into binary classification to facilitate a clearer, more straightforward interpretation of public opinions, and avoid the complexities of neutral or fine-grained sentiment analysis, which demands more nuanced approaches. As a result of our preprocessing, our method was tested on 446 textual comments from the EBLN survey, with 52 comments labelled as positive.

3.3 Service 1: Sentiment Classification

For the sentiment classification task in this study, we have tested Flan-T5 language model in three different model sizes (see Sect. 2.4).

To provide a comprehensive comparison, we tested these models both on a laptop device with an Intel Core i5 processor, and on an NVIDIA Titan RTX GPU-equipped computer. This additional test helped us quantify the reduction in execution time achievable with more powerful hardware, providing insights into the scalability of our approach.

All the Flan-T5 model variants mentioned were tested using identical prompt instructions, to ensure a consistent and fair comparison across different model sizes.

We aimed to use a clear and straightforward prompt, including only the essential components, namely the task name, task definition, and output format. Prompt format and wording were taken from [28] (Tables 3, 4 and 5).

In the prompt, we specify the name of the task (Sentiment Classification), and provide instructions on the task itself (assign a label to a given sentence). We also specify the output format we require after the final word 'Label'. We have executed this prompt for every data point in the dataset, replacing `{input_text}` with the appropriate text content from the dataset.

3.4 Service 2: Summarisation

For the purposes of this use case, we used Llama 3 with 8 billion parameters, and ran the model on an NVIDIA Titan RTX GPU-equipped computer.

To test the efficacy of the prompt engineering approach in the summarisation service, we experimented with two types of prompts; with context and without context. Prompts with context included additional information or requests provided by users. Prompts without context only included the actual system instructions, without any additional user instructions. Furthermore, we also experimented with two different system prompts. System Prompt 1 specified a more elaborate prompt with instructions to the LLM to be helpful. System Prompt 2 was more bare bones and only instructed the LLM to provide a summary. The two prompts used are shown below.

A streamlit webapp (Fig. 1) was created to allow different users to upload their own datasets and experiment with providing various contexts. These contexts could also include instructions for the LLMs to tailor the summary that was generated. For example, it could include instructions to "focus only on transport related responses", or "what do citizens say about bus services?".[9]

For each prompt, 9 different contexts were devised. Therefore, including no context, there were 10 total summaries generated. The summaries were then ranked by domain experts as to their usefulness from 1 to 5; 1 being least useful and 5 being most useful. The results are discussed in Sect. 4.2. The contexts are shown in Table 6.

---

[9] https://streamlit.io/.

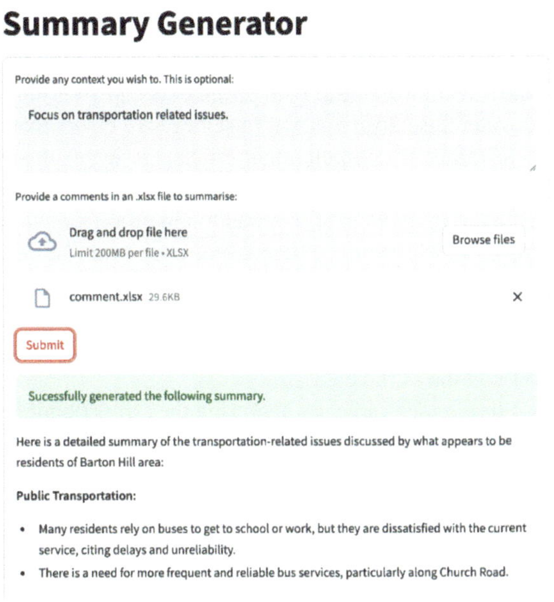

Fig. 1 A Streamlit webapp created to experiment with different contexts

Table 6 Contexts supplied by domain experts

ID	Context
1	No context
2	What do people think about bus services
3	What do people think about cycle lanes
4	Tell me about bike storage in the area
5	What do people like about their area
6	Tell me about green spaces
7	Tell me about air quality
8	What do people think about traffic
9	What do people think about anti-social behaviour
10	What is the biggest issue in the area

4 Results and Discussion

4.1 Sentiment Analysis

The performance of our approach is reported in Table 7. In our evaluation of the Flan-T5 models, we focused on two key classification metrics, accuracy and F1-score, since they respectively provide an overview of the overall correctness, and the balance between precision and recall. The accuracy results demonstrate a clear progression as model size increases. The Flan-T5-large model performs significantly better, with

92% of its predictions being correct, and there is a notable jump in accuracy from the base to the large model, in line with the increase in model parameters. The F1 scores follow a similar trend, with all models showing high scores, hence a good balance between false positives and false negative errors made.

In terms of computational efficiency, we evaluated the model's execution time on both CPU and GPU configurations. Running Flan-T5 with GPU acceleration results in roughly an 80% reduction in processing time compared to CPU execution. However, it is worth noticing that even with the larger model requiring > 14 min for processing, this is still substantially faster than the time an analyst would spend classifying such data manually, which, according to staff members at Bristol City Council, could range from a couple of hours to days depending on the complexity and volume of the data.

For Flan-T5-large, by analysing the errors obtained, we see in particular that only one positive instance gets predicted as negative, and we have 34 false positives. Looking at these errors, we can notice that they occur mostly on statements that are not fully negative or positive in the sentiment spectrum, or where mixed feelings are present in the same statement. Table 8 shows some examples of these errors, where 'True' corresponds to the label selected by the respondent and simplified to binary form during our data preprocessing phase.

This suggests that we need more nuance in our approach, perhaps by considering a bigger range of sentiments, or by analysing sentiments at sentence level or on different topics within each statement, rather than assigning a single sentiment to

Table 7 Performance metrics for different Flan-T5 models

Model	Accuracy	F1-score	Time (CPU)	Time (GPU)
Flan-T5-small	0.81	0.88	1m43s	20 s
Flan-T5-base	0.85	0.91	4m49s	50 s
Flan-T5-large	0.92	0.95	14m15s	2m25s

Table 8 Some false positives and negatives results from Flan-T5-large

Statement	True	Predicted
There should be trees planted at this junction, there is plenty of space on the central island and wide footway. This would make the area look more pleasant and less industrial. They would also help absorb pollution	Negative	Positive
Perfect place for diagonal pedestrian crossing please please please?! I can think of no better place for this in Bristol. So many stages to crossing this street. Phasing of lights should favour pedestrians and cyclists not cars	Negative	Positive
The public rights of way on Troopers Hill need diverting to follow the existing paths, at the moment OS and other maps show routes that can't be followed	Positive	Negative

the full statement. Such an approach could better capture the complexity and potential ambiguity in responses, particularly those containing multiple, not binary, or conflicting sentiments.

Overall, these results are quite positive. The consistent improvement in both metrics as model size increases suggests that larger models do indeed capture more nuanced patterns in the data, leading to better classification performance. However, this comes at the cost of increased execution time. Despite this, the computational time remains significantly lower than what would be required for manual analysis, demonstrating the efficiency of automated approaches even with larger models.

4.2 Summarisation

For summarisation, we performed a domain expert evaluation. The results are shown in Fig. 2. The most useful summary was generated without any additional context for both system prompts (*Context 1*). However, there is a clear divergence between the two system prompts for all contexts after that. Overall, the scores of the summaries were inconsistent. For example, for *System Prompt 1*, *Contexts 5* and *8* were rated higher than other summaries. For *System Prompt 2*, the scores of the summaries were significantly lower. This seems to confirm the findings of Işıkdemir [9] that summaries are highly sensitive to the prompts. A simple instruction like '*to be more helpful*' clearly improves the scores of the results, though the LLM appears unable to understand instructions to focus on specific topics or elements. Using a Retrieval-Augmented Generation (RAG) based approach may help alleviate this issue [11].

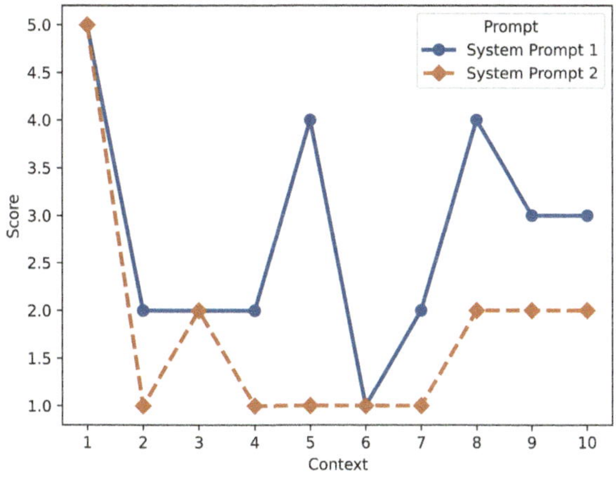

Fig. 2 Human evaluated scores for different contexts and different system prompts

5 Conclusion and Future Works

In this paper, we argued that Generative Artificial Intelligence (GAI), more specifically Large Language Models (LLMs) can be used for data analysis by public authorities. More specifically, we investigated sentiment analysis and summarization using open-source LLM models and applied them to survey data from the East Bristol Liveable Neighbourhood case study. The results show that larger versions of LLMs consistently improve accuracy in sentiment classification tasks. As model size increased, performance metrics showed steady improvements, indicating better capture of nuanced patterns. However, this came with longer execution times, although remaining more efficient than manual analysis. Summarisation quality varied significantly based on the prompts used: explicit instructions produced better summaries, while additional context did not consistently improve results, according to experts' evaluation.

We anticipate that such GAI-based services can provide efficient means to public authorities in performing data analysis and evaluating or assessing public opinion. These systems can process diverse data sources, including social media, comments on planning applications, petitions, etc. However, further research is needed to improve the accuracy of sentiment classification and summarisation of survey data. Our future work will focus on comparing LLMs with conventional models across these diverse data sources to understand their relative strengths and weaknesses. We plan to evaluate various LLMs for task-specific suitability, and explore prompt engineering techniques to optimise performance. For sentiment classification, we aim to implement Aspect-Based Sentiment Analysis (ABSA) to understand more nuanced patterns, and provide insights into both emotions and topics within the text. Regarding summarisation tasks, we intend to incorporate Retrieval-Augmented Generation (RAG) techniques to improve contextual relevance by implementing pre-filtering of responses before inputting to the LLM for summarisation. Additionally, we will investigate how biases inherent in the training data of LLMs can impact the results and performance of the models, as understanding these biases helps in ensuring fair and accurate results.

Acknowledgements This work is funded by the GREENGAGE project. [10] The project has received funding from the European Union's Horizon Europe research and innovation programme (grant agreement number 101086530). This work is co-funded by UK Research and Innovation (UKRI) under the UK government's Horizon Europe funding guarantee project reference numbers: 10063933, 10062587, 10064231. The project is also co-funded by SERI (State Secretariat for Education, Research and Innovation) under the funding ID Contract No. 22.00496 REF-131-52014.

Disclosure of Interests. The authors have no competing interests to declare that are relevant to the content of this article.

[10] The GREENGAGE project—https://www.greengage-project.eu/.

References

1. Bassoo, V., Ramnarain-Seetohul, V., Hurbungs, V., Fowdur, T.P., Beeharry, Y.: Big Data Analytics for Smart Cities, pp. 359–379. Springer International Publish- ing, Cham (2018). https://doi.org/10.1007/978-3-319-60435-0_15
2. Basyal, L., Sanghvi, M.: Text summarization using large language models: A com- parative study of mpt-7b-instruct, falcon-7b-instruct, and openai chat-gpt models (2023), https://arxiv.org/abs/2310.10449
3. Chang, M.W., Ratinov, L., Roth, D., Srikumar, V.: Importance of semantic rep- resentation: Dataless classification. In: Proceedings of the National Conference on Artificial Intelligence. vol. 2, pp. 830–835 (2008)
4. Chung, H.W., Hou, L., Longpre, S., Zoph, B., Tay, Y., Fedus, W., Li, Y., Wang, X., Dehghani, M., Brahma, S., Webson, A., Gu, S.S., Dai, Z., Suzgun, M., Chen, X., Chowdhery, A., Castro-Ros, A., Pellat, M., Robinson, K., Valter, D., Narang, S., Mishra, G., Yu, A., Zhao, V., Huang, Y., Dai, A., Yu, H., Petrov, S., Chi, E.H., Dean, J., Devlin, J., Roberts, A., Zhou, D., Le, Q.V., Wei, J.: Scaling instruction- finetuned language models. Journal of Machine Learning Research 25(70), 1–53 (2024), http://jmlr.org/papers/v25/23-0870.html
5. Covato, E., Jeawak, S.: Understanding People's Perceptions of Their Liveable Neighbourhoods: A Case Study of East Bristol. In: 12th International Confer- ence on Geographic Information Science (GIScience 2023). Leibniz International Proceedings in Informatics (LIPIcs), vol. 277, pp. 24:1–24:6 (2023). https://doi.org/10.4230/LIPIcs.GIScience.2023.24
6. Davies, S.R., Selin, C., Gano, G., Ângela Guimarães Pereira: Citizen engagement and urban change: Three case studies of material deliberation. Cities 29(6), 351– 357 (2012). https://doi.org/10.1016/j.cities.2011.11.012
7. Davoodian, Z., Mohebi, A.: Unveiling sentiment insights in smart cities: Exploring the role of social media. In: 2024 8th International Conference on Smart Cities, Internet of Things and Applications (SCIoT). pp. 68–75 (2024). https://doi.org/10.1109/SCIoT62588.2024.10570117
8. Halfpenny, P.: The analysis of qualitative data. The Sociological Review 27(4), 799–827 (1979)
9. Işıkdemir, Y.E.: NLP transformers: Analysis of LLMs and traditional approaches for enhanced text summarization. Eskişehir Osmangazi Üniversitesi Mühendislik ve Mimarlık Fakültesi Dergisi 32(1), 1140–1151 (2024). https://doi.org/10.31796/ogummf.1303569
10. Izzidien, A., Sargeant, H., Steffek, F.: LLM vs. Lawyers: Identifying a Subset of Summary Judgments in a Large UK Case Law Dataset (Mar 2024). https://doi.org/10.48550/arXiv.2403.04791
11. Jeong, C.: Generative AI service implementation using LLM application architec- ture: based on RAG model and LangChain framework. Journal of Intelligence and Information Systems 29(4), 129–164 (2023). https://doi.org/10.13088/jiis. 2023.29.4.129
12. Khan, Z., Ludlow, D., Loibl, W., Soomro, K.: Ict enabled participatory urban planning and policy development. Transforming Government: People, Process and Policy 8(2), 205–229 (Jan 2014). https://doi.org/10.1108/TG-09-2013-0030
13. Krugmann, J.O., Hartmann, J.: Sentiment analysis in the age of generative ai. Customer Needs and Solutions 11(1), 3 (Mar 2024). https://doi.org/10.1007/s40547-024-00143-416
14. Liu, P., Yuan, W., Fu, J., Jiang, Z., Hayashi, H., Neubig, G.: Pre-train, prompt, and predict: A systematic survey of prompting methods in natural language processing. ACM Comput. Surv. 55(9) (2023). https://doi.org/10.1145/3560815
15. Mohammadi, M., Al-Fuqaha, A., Sorour, S., Guizani, M.: Deep learning for iot big data and streaming analytics: A survey. IEEE Communications Surveys & Tuto- rials 20(4), 2923–2960 (2018). https://doi.org/10.1109/COMST.2018.2844341
16. Naveed, H., Khan, A.U., Qiu, S., Saqib, M., Anwar, S., Usman, M., Akhtar, N., Barnes, N., Mian, A.: A comprehensive overview of large language models (2024), https://arxiv.org/abs/2307.06435
17. Orlanski, G.: Evaluating prompts across multiple choice tasks in a zero-shot setting (2022), https://arxiv.org/abs/2203.15754

18. Poplin, A.: Playful public participation in urban planning: A case study for on- line serious games. Computers, Environment and Urban Systems 36(3), 195–206 (2012). https://doi.org/10.1016/j.compenvurbsys.2011.10.003
19. Rasheed, Z., Waseem, M., Ahmad, A., Kemell, K.K., Xiaofeng, W., Duc, A.N., Abrahamsson, P.: Can large language models serve as data analysts? a multi-agent assisted approach for qualitative data analysis (2024), https://arxiv.org/abs/2402.01386
20. Sarker, I.H.: Machine learning: Algorithms, real-world applications and research directions (2021). https://doi.org/10.1007/s42979-021-00592-x
21. Sharma, N.A., Ali, A.B.M.S., Kabir, M.A.: A review of sentiment analysis: tasks, applications, and deep learning techniques. International Journal of Data Science and Analysis (2024). https://doi.org/10.1007/s41060-024-00594-x
22. Soomro, K., Bhutta, M.N.M., Khan, Z., Tahir, M.A.: Smart city big data analytics: An advanced review. WIREs Data Mining and Knowledge Discovery 9(5), e1319 (2019). https://doi.org/10.1002/widm.1319
23. Tai, R.H., Bentley, L.R., Xia, X., Sitt, J.M., Fankhauser, S.C., Chicas-Mosier, A.M., Monteith, B.G.: An examination of the use of large language models to aid analysis of textual data. International Journal of Qualitative Methods 23, 16094069241231168 (2024). https://doi.org/10.1177/16094069241231168
24. Wang, Z., Pang, Y., Lin, Y.: Large language models are zero-shot text classifiers (2023), https://arxiv.org/abs/2312.01044
25. Xiao, Z., Yuan, X., Liao, Q.V., Abdelghani, R., Oudeyer, P.Y.: Supporting qual- itative analysis with large language models: Combining codebook with gpt-3 for deductive coding. In: Companion Proceedings of the 28th International Confer- ence on Intelligent User Interfaces. p. 75–78. IUI '23 Companion, Association for Computing Machinery, New York, NY, USA (2023). https://doi.org/10.1145/3581754.3584136
26. Zaouali, K., Rekik, R., Bouallegue, R.: Deep learning forecasting based on auto- lstm model for home solar power systems. In: 2018 IEEE 20th International Conference on High Performance Computing and Communications; IEEE 16th International Conference on Smart City; IEEE 4th International Conference on Data Science and Systems (HPCC/SmartCity/DSS). pp. 235–242 (2018). https: //doi.org/https://doi.org/10.1109/HPCC/SmartCity/DSS.2018.00062
27. Zhang, T., Ladhak, F., Durmus, E., Liang, P., McKeown, K., Hashimoto, T.B.: Benchmarking Large Language Models for News Summarization. Transactions of the Association for Computational Linguistics 12, 39–57 (01 2024). https://doi.org/10.1162/tacl_a_00632
28. Zhang, W., Deng, Y., Liu, B., Pan, S.J., Bing, L.: Sentiment analysis in the era of large language models: A reality check (2023), https://arxiv.org/abs/2305. 15005

GPSR Compliance

The European Union's (EU) General Product Safety Regulation (GPSR) is a set of rules that requires consumer products to be safe and our obligations to ensure this.

If you have any concerns about our products, you can contact us on ProductSafety@springernature.com

In case Publisher is established outside the EU, the EU authorized representative is:

Springer Nature Customer Service Center GmbH
Europaplatz 3
69115 Heidelberg, Germany

Batch number: 10755466

Printed by Printforce, the Netherlands